Cultural Dynamics of Water in Iranian Civilization

Majid Labbaf Khaneiki

Cultural Dynamics of Water in Iranian Civilization

Springer

Majid Labbaf Khaneiki
UNESCO ICQHS
Yazd, Iran

ISBN 978-3-030-58899-1 ISBN 978-3-030-58900-4 (eBook)
https://doi.org/10.1007/978-3-030-58900-4

This Springer imprint is published by the registered company Springer Nature Switzerland AG
The registered company address is: Gewerbestrasse 11, 6330 Cham, Switzerland

Foreword

Our history is awash with many pandemics that blighted many civilizations in the world, such as the pandemic that crippled both Athens and Sparta amid Peloponnesian War in 430 BC, or the plague of Justinian that swept away the Byzantine Empire between 541 and 750 AD. The modern history has not been spared from such pandemics, as the Spanish flu proved to be one of the deadliest calamities that afflicted human societies after the world war one. Even our recent scientific and technological breakthroughs could not have shielded us from such pandemics as COVID-19 that is rampaging through almost all cities and villages in the world claiming tens of thousands of lives, just when I am writing this foreword. Such historical experiences are so horrific that one may come to think about those pandemics as doomsday scenarios taking our species to the brink of extinction. But nothing seems to have been able to scale down the vanity of our modern civilization, even another silent pandemic that dwarfs Coronavirus.

Climate change is creeping into every aspect of our social life, and inflicted irreversible damages on a vast area from the North Pole to Antarctica, from as tiny as a phytoplankton to as big as a blue whale. This pandemic is transmittable not only to every individual but also to every creature and every landscape and is looming larger and larger every year. All scientific estimates show that this pandemic would be more devastating than any pandemic that humankind has ever gone through over the course of history. That is why the Swedish non-profit Global Challenges Foundation has ranked climate change third among the deadliest threats to humanity after nuclear wars and chemical-biological attacks, whereas a pandemic like COVID-19 has been ranked fifth.

In the wake of a pandemic, all scientists and laboratories from across the globe bend over backward to find an urgent remedy in order to make humans invincible against the new diseases. Our modern medical advancements are in fact indebted to such diseases that spring up out of the blue once in a while. Of course, the scientists have not been apathetic about climate change either, and they have come up with two key solutions, though their recommendations have sometimes fallen on deaf ears. According to United Nations Framework Convention on Climate Change (UNFCCC), the possible solutions consist in climate change mitigation by reducing or halting emissions of greenhouse gases, and enhancing our adaptation to climate change effects.

The first solution is highly contingent on an intergovernmental consensus to mobilize all resources and possibilities of the world, a great program whose first step was taken at the 2015 United Nations Climate Change Conference in Paris. However, the second solution is in our own hands and we can easily look around to learn from our geography and history how to live in harmony with nature. The way our ancestors could live side by side with their natural environment still holds precious lessons for us to realize the second solution; adaptation. Although historical communities did not have to cope with anthropogenic climate change on such a gigantic scale, they have never been spared from climatic fluctuations since the dawn of history. Indigenous people could have built up a great deal of knowledge and experience that can be a source of inspiration for us to better adapt to our changing climate. We should ponder how ancient civilizations could flourish and sustain their populations in the face of natural limitations, how they could keep their economic systems in balance with their environmental possibilities, and how they could re-adapt their socio-economic structures to their changing climatic conditions. Now it is time for us to dust the history book of our ancestors, but this time to look up the secret of survival not the matter of pride and chauvinism.

Our ancestors could exhibit a high level of adaptation to their harsh environment where water scarcity has always been a major hindrance to human civilizations. Nevertheless, their efforts for adaptation went far beyond a simple struggle for existence, but it turned out to be a basis for a prosperous civilization that was coveted for centuries by many even those who lived in favorable geographies. Water scarcity gave rise to a culture of adaptation that has been manifest in every aspect of the Iranians' social life. It is a must for us to delve further into the culture of adaptation that has protected our ancestors against many environmental ups and downs in their history, and this book is one of such efforts.

This book presents the result of years of research conducted by Dr. Majid Labbaf Khaneiki, senior researcher of UNESCO-ICQHS. In this volume, he sheds light on the Iranians' culture of adaptation that has been acquired through centuries of water management. This book provides a set of historical and anthropological information, which portrays a different dimension of water in Iranian culture and civilization. Water is traced back to the farthest corners of Iranian mind, political economy and social atmosphere from a historical standpoint. This volume can be a helpful reference for those who wish to build upon history in order to reconfigure a resilient sustainable water management.

Mehdi Javadian-Zadeh, Ph.D.
Director of International Center on
Qanats and Historic Hydraulic
Structures (UNESCO-ICQHS)
Yazd, Iran

Acknowledgements

The idea for this book grew out of an earlier work on territorial water cooperation in the central plateau of Iran. The research that resulted in this book was funded by the International Center on Qanats and Historic Hydraulic Structures (UNESCO-ICQHS). I owe a great debt of gratitude to Dr. Mehdi Javadian-Zadeh, ICQHS director, for his considerable support in bringing this work to fruition. I sincerely thank all my ICQHS colleagues without whose help this book would not have been written, particularly to Ali Bahri, Soudeh Semsar Yazdi, Seyyed Mohammad Jalili and Hossein Dehghani.

I owe a particular debt to my daughter Kiana who did the illustrations for the book, and to my wife, Pari whose unfailing encouragement paved the way for me to accomplish this work. I also acknowledge my brother Meysam and my father whose comments have been essential to the making of this book. Eventually I heartily thank my dear mother Ashraf Banoo Khaniki who has always been a role model for me, showing me that human's strength and stamina are infinite. I would welcome correspondence from dear readers, which would be of great help to me to rectify errors or omissions in future editions of the book.

Contents

Chapter 1
Introduction

Abstract Water is the only chemical substance whose footprint is omnipresent throughout the Iranian culture. Water has been so deeply engraved on Iranian culture that it can be traced back to the furthest corners of their dreams and delusions. Water has also leaked into their philosophical minds attempting to find an explanation for the existence of universe, as described in Zoroastrian Akhshij philosophy according to which water is one of the four fundamental elements of the creation. This book goes beyond indigenous water knowledge and technologies, but water will be conceptualized as a pivotal element of Iran's social identity, cultural dynamics and belief systems. The book tries to shed light on how the cultural structures of the Iranian local communities have been inundated by "water". A systematic analysis is deployed to discover a mesh of relationships between four principal factors—"water", "society", "culture" and "natural environment"—in a dynamic historical context. I have tried not to remain limited to a myopic causal relation between water and society, nor do I wish to reduce my argument to a deterministic relation between natural environment and culture. Instead, I have attempted to factor in a wider range of determinants in a moving background called "history".

1.1 The Concept of Water Culture

Our human cognition is so overwhelmed by "culture" that it is almost impossible to divest our philosophy on culture of cultural influences. It is difficult for us to come up with a precise definition of culture, the way we cannot figure out the genuine boundary of the universe surrounding us. Human nature gave way to culture that gradually became omnipresent in every aspect of our notions and actions. Our ancestors invented culture that set them apart from other mammals, by flouting the laws of nature. Culture is the fruit of human self-awareness, which was reflected for the first time in the prehistoric cave drawings, human or animal figurines, beads or other ornaments. Those humans with such cognitive skills could not only mastermind a set of elaborate tools and group activities, but could also express their fears,

M. Labbaf Khaneiki, *Cultural Dynamics of Water in Iranian Civilization*,
https://doi.org/10.1007/978-3-030-58900-4_1

imaginations, hopes and admirations for their place in the intricacies of nature (Klein and Edgar 2002, p. 14).

Culture enhanced our ancestors' capability of adapting to environmental change. If humans left their adaptation only to their natural evolution, they probably went down the same path of many extinct species whose natural evolution could not keep up with the pace of environmental change. This strategy of cultural adaptation helped humans gain the upper hand over the other species and break the law of the jungle, as the dominant life on the earth (Ibid, p. 26). Hence, we can say that culture helped the primitive humans break free from spatial and temporal limitations that underlay their natural evolution. Prior to culture, humans were bound up with space and time as two pillars of their biological evolution. Natural spaces and their differentiation paved the way for genetic diversity and then natural selection, albeit in the context of time. According to Heraclitus (535–475 BC), nothing in the world can exceed its measures that are spatial and temporal limitations (Cassirer 1944, p. 62). However, humans did so through culture. In the absence of culture, humans were not very different from sea turtles, for example, whose behaviors are totally molded by space and time. The Hawksbill sea turtle migrates thousands of kilometers to reach Kish Island in the Persian Gulf in order to lay her eggs. Every year, the turtle shows up in early spring, dragging herself onto the sandy beach at night. She leaves her eggs buried under the sand and immediately returns to the sea to start another round of her life cycle. Just as the eggs hatch, the baby turtles crawl into the sea according to the spatial and temporal measures programmed in their genomes. The turtle does not teach her babies where and when to go after hatching out, and where and when to lay eggs in the future after mating. They are all confined to the natural boundaries of space and time, which are inherent in their genomes, in compliance with environmental conditions. However, when a human baby is born, she has no perception of space and time at birth. She has to learn thousands of wheres and whens, which gradually shape her cultural atmosphere. Human geography has transcended the natural concepts of space and time. A turtle goes to different places by instinct. And each place actually forms in front of her eyes as soon as she arrives, and then disappears as soon as she leaves. None of her places is connected to any past and future, or carries any special meaning, philosophy and function existing independent of space and time. However, when a human sets foot in a school for the first time, her experience of arriving in the school does not remain limited to the moment of her arrival, but it evokes her past conversations with her parents about school, many stories, memories and movies, and it spreads to many dreams, ideas and future prospects. Her school protrudes from the framework of the time and space of her arrival, and annexes many conceptual territories that are timeless and placeless like education, development and knowledge. Humans even created new spaces and times, which had no physical manifestation in their geographical environment. Heaven and genesis cannot be traced on any physical map and calendar, but they still played a crucial role in human history and geography.

When humans stepped out of the natural boundaries of space and time, they entered a dynamic territory called culture. Every mammal lives in a standard ecological niche, whereas humans are spread out almost everywhere on this planet, from the North Pole to Antarctica. They have gained the capability to break free from space and time through their culture. Inuit gave a new identity to their space by inventing a myriad of tools and techniques, in fact through their cultural adaptation to such a frigid environment that naturally did not favor human existence at all. The same story unfolded on the Iranian plateau where the desert's water scarcity and harsh environment do not fit in with the humans' physical capabilities bestowed on them by their biological evolution. They had to step out of their arid space in order to survive their thirst and hunger. Thus, they developed a dynamic water culture that established a new balance between their communities and their natural environment. In the heart of the desert, thousands of water streams sprang up and a myriad of green gardens appeared all over the landscape, which defied all spatial and temporal limitations and even expanded into the placeless Garden of Eden and the timeless moment of the creation. Hence, water became part of an inherently shared, cross-cultural experience (Rogers 2018, p. 2).

Altogether water is the most abundant element on the earth and is also the most serious problem that humans have ever faced in the course of their history, a problem that used to overwhelm humans because of whether its influx or its scarcity. We set foot in the world with plenty of water that makes up about 80% of our body as infant. Somewhere between 70 and 75% of the earth's surface is covered with water. Hence, we are surrounded all the way by water in our environment and we are also awash with water inside our own bodies.

For both purposes—preservation of water for humans and protection of humans against water hazards—the ancient people came to invent a variety of technologies and systems, each of which left a deep historical impact on human culture. Hence, water is the only chemical substance whose footprint is visible throughout our culture. Almost every culture used to worship a water deity whose duty was to protect water supplies for humans or to safeguard the humans against water hazards or both. Our human cultures have been laden with water goddesses, gods or spirits with different names, which imply the importance of water in human mentality, like Chalchiuhtlicue in Aztec Mexico (Dehouve 2020), Acionna in Celtic Culture (MacGulloch 1911, p. 182), Mazu in China (Ruitenbeek 1999), Tefnut in Egypt (Pehal and Svobodová 2018, p. 116), Halig-kelda in Scandinavia (Varner 2009, p. 156), Acheloös in Greece (Lee 2006), Apam Napat in India (Findly 1979), Pariacaca in Incan culture (Spence 1926, p. 406), Suijin in Japan (Kalland 1995, p. 177), Abzu in Mesopotamia (Albright 1919, p. 177), Tiberinus in Rome (Lipka 2009, p. 90) and eventually Anahita in Persia. Anahita was believed to be "as great as all waters running on the earth, owning one thousand lakes and one thousand rivers" (Pourdāvoud 1977, p. 281). Anahita's divine control over waters bestowed on her so great power that some Sassanian kings preferred only Anahita to lend legitimacy to their reign (see Fig. 1.1) (Alamolhodayi 2015, p. 111).

Water has been engraved on Iranian culture so deeply that it can be traced back to the mysterious seclusion of the Sufi monks. For example, Ibn Arabi (1165–1240 AD) gave a metaphorical meaning to "water" in order to clarify his cosmological teachings that had a great influence on the Iranian Sufism for centuries. He used this metaphor to characterize the notion of multiplicity in the face of the "Unity of

Fig. 1.1 Naqsh-e Rostam rock relief showing Narseh (228–303 AD) seventh king of the Sassanian Empire, who is receiving the sign of power from Anahita goddess of water, at his coronation (Gyselen 2005, p. 33)

Existence". According to his school of thought, water exemplifies the pure existence that is intrinsically united but at the same time manifested in such multiple forms as drops, sea, wave, river, moisture, rain, cloud, dew, hail, snow and ice. And multiple manifestations of water—as a metaphor for the pure existence—lead to the creation of various characteristics like purity, fluidity, necessity, omnipresence, freshness, shapelessness, transparency, and so on. Many Persian poets and philosophers later echoed this notion that became the cornerstone of their mystical interpretation (Heidari 2018; Zowghi 2012). Nonetheless, many believe that such metaphorical interpretations can be traced back to Zoroastrianism whose marriage with Islam gave birth to Persian Islamic Sufism (Sepahvandi 2008).

Water leaked into the philosophical minds attempting to find an explanation for the existence of universe, as described in Zoroastrian Akhshij philosophy according to which water is one of the four fundamental elements of the creation. Zoroaster believed that the entire universe was created from water (Abbasi et al. 2017, p. 118), even before the other fundamental elements. Ahura-Mazda first created water from a drop of his tear, and then created the sky to separate the earthly waters from the celestial ones (Ibid, p. 120). Hence, water deserved to be praised and even worshiped (Pourdāvoud 1964). In Zand-i Wahman Yasn, a Zoroastrian apocalyptical text, one of the Zoroaster's revelations is described as follows. "Ahura-Mazda (supreme god) handed over the divine omniscience to Zoroaster in the shape of water. Zoroaster got the water and took a sip of it, and suddenly all the world was revealed to him. He could see even every strand of hair of every single person, every leaf of every tree, and every root underground" (Moarefy 2012, p. 331).

Water was believed to harbor a secret knowledge that could be disclosed only to pious people. That was why water was used for trial by ordeal. This custom is reflected in the Laws of Hammurabi too. The ordeal of water was common to judge serious cases such as sorcery (Wright 2009, p. 200) and adultery (Ibid, p. 257) accusations. The accused was ordered to plunge into a river whose divine soul could shed light on the truth. If the accused was submerged and drowned, the accuser was legally allowed to expropriate the offender's properties. If the accused survived the ordeal of water, she/he was acquitted, and instead the accuser was put to death and her/his properties were confiscated for the accused's benefit.

This book goes beyond indigenous water knowledge, distribution of water, equal access to clean water, and so on, but water will be conceptualized as a pivotal element of Iran's social identity, cultural dynamics and belief systems. This book tries to shed light on how water could have overwhelmed the cultural structures of the local communities in Iran. Iran enjoys various climatic and geographical conditions, which have led to different strategies that the inhabitants have taken to adapt themselves to their environments. In the central plateau of Iran, the climatic and geographical conditions gave rise to the technology of qanat that underlay the production systems in this region. There is a rich literature about the environmental, technical, cultural and historical values of qanat, but one cannot find many references on the role of qanat in the cultural identities of its beneficiary societies. Qanat could have built up a high capacity for cooperation, social convergence and cultural dynamics, which can be used in configuring sustainable rural development. This book tracks down the footprint of water in different aspects of human culture in the Iranian arid and semi-arid lands, and it also examines the role of drought in engendering and spreading intangible cultural elements.

In Iran, water has established a special relationship between people and their environment, a relationship that underlies an intricate nexus of political, social, cultural and economic structures. This nexus is manifest not only in the adobe architecture, handicrafts, traditional bazaars, orchards, and so on but also in many intangible cultural elements such as cooperative systems, customs and beliefs. Watercourses like qanats are not just some technical objects aloof and separable from rural and urban life. In contrast, qanats are like a network of veins spreading all over the body of the city to bring life and prosperity everywhere, creating a special cultural landscape. The novelty of this book is down to its focus on the footprint of water in such intangible aspects of our culture.

Since the dawn of Iranian history, water has been at the center of Iranians' attention, giving rise to thousands of technologies, philosophies, arts, and so on. In this book, I leafed through hundreds of pages written about water in search of its cultural influence to better understand why water always stands out not only as matter of life but also as driver of culture in Iran. For example, in many desert towns, water resources are interwoven with the urban fabric in which waqf or charitable endowment plays a pivotal role. The rich tend to donate hydraulic structures such as qanat, water reservoir and pool in order to dampen the conflict inherent in the economic relationships between social classes.

Water culture also extends to the other realms of social life such as religion and mythology. For example, in Yazd province, in the village of Qasem Abad majority of whose inhabitants are Zoroastrian, the footprints of water are visible in every aspect of their cultural life. The practice of polluting water for any reason is abhorrent to the villagers. The village women are allowed to do the laundry by the qanat flow only at definite spots in the village, but they cannot dump the filthy water directly into the flow. There are some septic wells next to the laundry spots, where the women can dispose off their sewage. Nonetheless, any kind of washing is banned—using the qanat flow—six days a month, according to the Zoroastrian calendar. They do not use our 7-day week to keep track of the cycle of days, but they have only three calendar units: day, month and year. Hence, they name each day of the month after a divine entity to address an event in their calendar. For example, Zoroaster's birthday is on the day of Khordād in the month of Farvardin. This type of calendar can be traced back to the Persian society before the advent of Islam (Gharib 1997, p. 18; Gray 1904). In Qasem Abad, people are forbidden to approach the qanat flow for washing every month on the days number 1, 6, 13, 19, 22 and 29, respectively, called Urmazd, Khordād, Tir, Farvardin, Bād and Mehr-Espand. The above-mentioned days got their names from the divine entities that are somehow related to water. Urmazd or Ahura-Mazda is the supreme creator who brought everything including water into existence. Khordād or Haurvatāt assisted Urmazd in creating water, and then became its guardian (Boyce 1985, p. 27). Tir or Tishtrya is the god of rain (Goharpour and Jalalkamali 2016, p. 1333). Farvardin got its name from a group of powerful supernatural beings called Farvashi or souls of the dead. When it rains, they vie with each other to obtain more water for their own families, settlements, tribes and countries (Boyce 2001). Bād or Vayu literally means wind, who is a deity responsible for moving the clouds to bring rain (Bakhtiyari 2011, p. 53). And eventually Mehr-Espand or Mahraspand is another deity believed to serve as the guardian of water alongside Khordād (Aghazadeh 2013, p. 8).

Some of such beliefs were devised to justify social inequality regarding access to water resources. According to the Iranians' ancient beliefs, Tishtrya makes rain by taking water from the legendary sea of Farākh-kard. But he cannot do so without the assistance of the souls of Ashavanān[1] (Jafari Dehaghi 2010, p. 30) 99999 of whom are guarding the sea of Farākh-kard (Kooshan 2001, p. 70). Avesta, the religious text of Zoroastrians, places a high value on Ashavanān, where it reports a conversation between Ahura-Mazda (God) and the prophet Zoroaster. Ahura-Mazda tells Zoroaster that nothing in the world could come into existence in the absence of Ashavanān's assistance. "It is because of Ashavanān's blessing that water keeps flowing out of the springs" (Doost-khah 1992, p. 407). "You should venerate the souls of those almighty Ashavanān who directed the God-created waters into the beautiful canals, otherwise those waters remained stagnant long after their creation" (Ibid, p. 417). Who are Ashavanān indeed? Ashavanān literally

[1]Ashavanān is the plural of Ashavan.

means righteous people whether dead or living (Ibid, p. 405). However, almighty Ashavanān refers to the religious instructors and saviors among whom the living ones enjoy stronger souls (Ibid, p. 408). Hence, the commoners always owe a debt of gratitude to the religious instructors (probably priests) and saviors (probably kings) for their almighty souls that bring water out of the springs and keep water running in the canals. Thus, Ashavanān's authority over water resources is deemed to be inherent in the divine creation of water, according to Zoroastrian theology.

In many arid and semi-arid lands, human settlements flourished around water resources that were limited compared to the growing water demand. Hence, it seems difficult to do justice in terms of water division among the consumers, and in many regions, social stratification correlates with the level of access to water resources. Many examples show that the more a particular social class enjoyed political and economic power, the better chance they had to get access to sufficient water supply in quantity and quality. Power relations played a crucial role in prioritizing the country's regions, when it came to allocating the available water resources. There was a complicated interplay between the social origin of the ruling power, ethno-ideological hegemony and geopolitical concerns, which drew invisible borders between the regions, giving rise to inequitable distribution of water resources across Iran. People were not suffering from "absence of water" but from "thirst" that has been foisted on them by power relations.

1.2 The Book Outline

In this book, a wide variety of issues from production to socio-political organizations to rituals has been subsumed under the title of cultural dynamics of water. This variety stems from the very nature of culture that has fully saturated our social being. "Culture refers not only to the music, literature, and arts of a society but also to all the other features of its way of life: prevailing modes of dress, routine living habits, food preferences, the architecture of houses and public buildings, the layout of fields and farms, systems of education, government, and law" (Fouberg et al. 2009, p. 29).

Chapter 2 demonstrates the importance of water culture in relation to production systems. A village in western Iran has been selected as a case study, where the qanat's water division system has remained almost intact, in the face of a socio-economic transition that took place in the village over the past few decades. Until 1962, a traditional feudal system was in place, according to which the qanat water was divided among 11 agricultural units on an 11-day irrigation cycle that is still in use, though the feudal system itself is long gone. This chapter investigates water in relation to other production factors, which formed a specific feudalism in the village. In Iranian villages, feudalism was not identical in character, but it varied from region to region based on the relationship between the lords and their peasants. The relationship that underlay the entire agricultural system was contingent on five production factors: workforce, plough, seed, land and eventually water. The more the production factors in the peasants' possession, the higher the status they could gain in the village's power relation. And water as the most important production factor played a crucial role in such a socio-economic equation, to the extent that we may call that kind of feudalism a dynamic hydro-feudalism. That is why we cannot evaluate Iranian feudal system as a fixed ubiquitous phenomenon, but smaller components and their interconnections gave rise to a feudal system that was specific to a particular region, and cannot be extrapolated to other regions.

Chapter 3 examines the systematic relationship between water and political power from a historical-geographical point of view. This relationship took different forms in river-based and qanat-based economies. In a river-based economy like that of Sassanian Empire, governments had a systematic dependence on a massive production surplus emanating from irrigated cultivation. Hence, governments used both bureaucratic and military tools to secure their control over water as the vital sources of their revenue. However, governments' interests were not limited to water only as source of finance, but they were also concerned about water to strengthen their political ties with their preferred territories through water allocation systems. Within their kingdoms, some hydro-social territories were given primacy in terms of water shares, through both intensive hydraulic constructions and water management systems. Hydro-social borders were delineated by the political organization through water allocation based on political, religious and social priorities, and those borders altered from time to time in the wake of the political changes. This chapter also shows how "water management" could set the stage for the collapse of those political organizations, when it outgrew the governments' internal capacities. However, in a qanat-based economy, a relationship between water and managerial systems did not lead to the accumulation of power in the shape of gigantic political systems. The qanats were spread out across the Iranian plateau where they tended to clump together in numerous sporadic clusters in different distant territories. None of those oases could not provide a production surplus alone, necessary to support an extended political system. Hence, the scattering of qanats resulted in the dispersed political powers, which did not favor the formation of a centralized government. The qanat-based economy paved the way for a cooperative bottom-up management system, which does not tally with the theories ever developed on "water-state relation" in the Middle East.

Chapter 4 pertains to hydro-social cohesion in the local communities on the Iranian plateau. Hydro-social cohesion has been evolved in the course of history as an adaptation strategy. Their collective response to the changing water resources were embodied in both water management systems and water-based social mechanisms. This chapter demonstrates how water affects the make-up of a local community through a variety of water-related social mechanisms like cooperation, qanat maintenance, division of labor and waqf (charitable endowments). There is a distinction between two types of social cohesion: "inner-territorial cohesion" and "trans-territorial cohesion", which are both associated with the geographical peculiarities of the water resources of a particular region. Inner-territorial cohesion pertains to a social bond that forms among the residents within a particular territory mainly based on their common interest in terms of their water resources, whereas trans-territorial cohesion refers to a social bond between different territories or neighboring communities within the hinterland of a particular qanat sharing the same water resource.

Chapter 5 examines an environmental-cultural cycle that can be called "drought pump" whose social and cultural mechanism has contributed to Iran's cultural dynamics. In the existing studies, the role of water in creating culture has been much discussed, but almost no study views water scarcity or drought as a driver of culture. According to historical records, over the past centuries, intermittent droughts in the central plateau of Iran played a crucial role in the cultural dynamics. The dry periods pushed the population surplus out of those regions when the environmental pressure went beyond the tolerance of their production systems. The local communities responded to climate change with different strategies, proportional to the magnitude of water scarcity. During wetter periods, they expanded their cultivated lands with a wider variety of crops and they made less lands lie fallow. When a year turned dry, they shrank the area of farmlands and resorted to a cropping pattern with lower water demand, and modified their water division system. During severe droughts, they put more focus on handicrafts and local industries in order to supplement their fickle agricultural economy. When the aridness crossed the tolerance line of the society to the point that even the offset economy did not help, part of the population began to migrate to more favorable regions, taking and spreading their cultural elements with themselves. When a more humid period followed the drought, the remaining population gradually returned to agriculture, which brought about more food production, and as a result population started to grow again and this cycle continued.

Chapter 6 traces the footprint of water in the Iranians' cultural mentality by searching through some rituals and customs in the rural communities. This chapter recounts a number of rain-claiming ceremonies that are expressive of the cosmology of the local people on the one hand, and of their relationship with water resources on the other hand. All those rain-claiming ceremonies fall into four categories: defiance, supplication, appeasement and combined actions. The notion behind all ceremonies is an omnipotent phenomenon like a supernatural master that mysteriously wields power over rain. Local people may take different strategies to align the rain master with their own needs. Sometimes locals consider the rain as a right that has been withheld from them at the time of drought. Hence, they do something weird in defiance of social, moral or even religious norms in order to put pressure on the rain master to make him relent and give the rain back to them. Sometimes they entreat the rain master to have mercy on them and compassionately quench their thirst, or they may appease him by throwing him a sop. In some other ceremonies, defiance, supplication and appeasement may be practiced in combination. This chapter examines "qanat marriage" and its socio-economic function as well. In some regions of central Iran, the locals used to throw a wedding party for their qanat (water gallery) when it came to dry up. They asked a widow in the village to become the wife of qanat and dwell in a house built just close to the qanat exit. The villagers treated the qanat's wife as a real bride, and all the customs were observed the same as that of a real wedding. Qanat marriage seems to be a holdover from the human's first attempt to understand the world by interpreting things, which is known as animism and anthropomorphism.

1.3 Research Method

The qualitative method of social sciences has been deployed in this book that deals with various societal and cultural factors in a traditional context. Qualitative research involves the gleaning, analyzing and interpreting of rich narrative information by looking into the beliefs, values and behaviors of the local communities in their own original social context. Given the nature of this research, I have opted for hermeneutic and interpretative approaches. In a sense, hermeneutic means scrutinizing the historical books, articles and documents regarding the concept of "water and cultural dynamics", and studying the situation of qanats and other water resources in the Iranian communities in order to get an accurate view of water and its position in the social-cultural fabric. Systematic analysis helped me better understand the subtle relationships between water and some cultural phenomena in their genuine environment. This approach tries to penetrate the visible layers toward the ultimate meanings by interpreting the connections and interactions. The local communities themselves, the historical and cultural data are all subject to this interpretation. Although this research deployed an interdisciplinary and holistic approach, it actually unfolded into two stages as follows: historical and geographical, each of which entailed specific methods in keeping with the nature of my available data.

In my historical studies, the data and information mostly came from thousands of pages of classical history records in Persian and Arabic languages, recent historiographies and books in different languages mostly in English, as well as numerous ethnographical reports. I tried to read through the relevant documents from a deductive standpoint, without any preconception that could distort the real cultural patterns.

In my geographical studies, I traveled to many local communities in order to find out about their socio-cultural structures in relation to their water resources. I preferred the qualitative method of grounded theory, which fitted in with the deductive method of the previous stage. This method deals with the human complicated interactions in a chronological background. Hence, this method focuses on each cross-section of time as the continuation of history, rooted in and affected by history (Streubert and Carpenter 2007). Grounded theory is made up of two interacting procedures: gleaning information and analysis. Sorting and coding the information were used as effective tools to facilitate the final analysis. Therefore, grounded theory is in fact a process through which the final theories gradually take shape and appear (Corbin and Strauss 2008).

My information fall into three categories: information about actors, information about events and information about conditions. The human and non-human actors are the main players like the villagers, their water resources, and so on. The events pertain to the relationships that take place between the actors in a traditional context. The conditions prevail over both actors and events like migration surges, governmental policies, and so on. All the required information were gleaned through three main sources: recorded documents, interviews and observations. The relevant documents were obtained by sifting through many books, articles and reports. Interviewing was an important tool leading me to the roots of the villages' intangible culture. I had many ethnographic interviews with the locals who were selected through a snowball sampling. Moreover, observation was deployed to get a view about the cultural waterscape, and to discover more subtle elements, which were mentioned neither in the recorded documents nor in the interviews.

References

Abbasi, S., Safari, M., & Asgari, L. (2017). Kārkard-hāye Ostoore-shenākhti-ye Āb Dar Irān-e Bāstān [Mythological functions of water in Ancient Iran]. *Āmoozesh-e Mohit-e Zist Va Towse'e-ye Pāydār, 6*(2), 117–126.

Aghazadeh, R. (2013). *Pirāmoon-e Nām-e Āturpāt Mānespandān [About the name of Āturpāt Mānespandān].* Zanjan: Anjoman-e Tarvij-e Zabān Va Adab-e Fārsi.

Alamolhodayi, P. (2015). Barresi-ye Tahavolāt-e Siyāsi-ye Dākheli-ye Hokoomat-e Sāsāniān Bā Tekyeh Bar Naghsh Barjaste-hā Va Katibe-hā [Internal political upheavals of the Sassanian Government with a focus on rock reliefs and inscriptions]. *Tārikh-nāmeye Khārazmi, 3*(2015), 104–120.

Albright, W. F. (1919). The mouth of the rivers. *The American Journal of Semitic Languages and Literatures, 35*(4), 161–195.

Bakhtiyari, R. (2011). Mazāmin-e Asātiri Dar Karde-ye Yekom-e Rām-Yasht [Mythological concepts in the book Rām-Yasht]. *Adabiyāt-e Erfāni Va Ostooreh-Shenāsi, 6*(21), 45–63.

Boyce, M. (1985). ĀB: The concept of water in Ancient Iranian culture. *Encyclopedia Iranica* (Vol. 1). Routledge.

Boyce, M. (2001). Fravaši. *Encyclopedia Iranica* (Vol. X, pp. 195–199). Routledge.

Cassirer, E. (1944). *An introduction to a philosophy of human culture.* Yale University Press.

Corbin, J., & Strauss, A. L. (2008). *Basics of qualitative research: grounded theory procedures and techniques.* Sage.

Dehouve, D. (2020). The rules of construction of an Aztec deity: chalchiuhtlicue, the goddess of water. *Ancient Mesoamerica, 31*(2020), 7–28.

Doost-khah, J. (1992). *Avestā: Kohan-tarin Soroode-hā Va Matn-hāye Irāni [Avesta: the Ancient Iranian hymns and texts].* Tehran: Morvārid.

Findly, E. B. (1979). The child of the waters: a revaluation of Vedic Apam Napat. *Numen, 26*(2), 164–184.

Fouberg, E. H., Murphy, A. B., & de Blij, H. J. (2009). *Human geography: people, place, and culture.* Hoboken: Wiley.

Gharib, B. (1997). Hafteh Dar Irān-e Ghadim [Week in the Ancient Iran]. *Nāmeh-ye Farhangestān, 12*(1997), 11–39.

Goharpour, H., & Jalalkamali, V. (2016). Studying the myth of water in Ancient Iran and its relationship with rain-making rituals. *The Turkish Online Journal of Design, Art and Communication.* https://doi.org/10.7456/1060AGSE/017.

Gray, L. H. (1904). The origin of the names of the Avesta months. *The American Journal of Semitic Languages and Literatures, 20*(3), 194–201.
Gyselen, R. (2005). Vahrām III (293) and the rock relief of Naqsh-i Rustam II: a contribution to the iconography of Sasanian Crown princes in the third century. *Iranian and Zoroastrian Studies in Honor of Prods Oktor Skjærvø, 19*(2005), 29–36.
Heidari, M. (2018). Āb Va Ashkāl-e Goonāgoon-e Ān Dar Adabiyāt-e Erfāni-ye Fārsi [Water and its various forms in Persian mystical literature]. *Matn-Pajoohi Adabi*. https://doi.org/10.22054/ltr.2018.33543.2320.
Jafari Dehaghi, M. (2010). Gozāresh-e Manoochehr Javān-Jam Az Tishtar-e Bārān-Saz Dar Matn-e Fārsi-ye Miyāne-ye Dādestān-e Dini [Manoochehr's report on rain-making Tishtrya according to the Middle Persian text of Dadestan-e Dini]. *Zabān-Shenākht, 1*(2), 27–38.
Kalland, A. (1995). Facing the spirits: illness and healing in a Japanese community. In M. Kõiva & K. Vassiljeva (Eds.), *Folk belief today* (pp. 167–182). Tartu: Estonian Academy of Sciences.
Klein, R. G., & Edgar, B. (2002). *The dawn of human culture*. New York: Wiley.
Kooshan, M. (2001). *Jostejooy-e Kherad-e Irāni [In search of Iranian wisdom]*. UK: H&S Media Ltd.
Lee, M. M. (2006). Acheloös peplophoros: a lost statuette of a river god in feminine dress. Hesperia: The Journal of the American School of Classical Studies at Athens, *75*(3), 317–325.
Lipka, M. (2009). *Roman gods: a conceptual approach*. Leiden: Koninklijke Brill NV.
MacGulloch, J. A. (1911). *The religion of the Ancient Celts*. Edinburgh: T. & T. Clark.
Moarefy, A. (2012). *Jahān-bini-ye Irāni [Iranian worldview: a comparative research in common denominators of Iranian myth, philosophy, and sufism with the new findings in sciences]*. Los Angeles: Sherkat-e Ketāb.
Pehal, M., & Svobodová, M. P. (2018). Death and the right fluids: perspectives from egyptology and anthropology. *Journal of Ancient Egyptian Interconnections, 17*(2018), 114–136.
Pourdāvoud, E. (1964). *Visperad*. Tehran: Asātir.
Pourdāvoud, E. (1977). *Yasht-hā [Zoroastrian hymns]*. Tehran: Daneshgāh-e Tehrān.
Rogers, D. K. (2018). *Water culture in Roman society*. Leiden: Brill.
Ruitenbeek, K. (1999). Mazu, the patroness of sailors in Chinese pictorial art. *Artibus Asiae, 58*(3/4), 281–329. https://doi.org/10.2307/3250021.
Sepahvandi, M. (2008). Moroori Bar Mafhoom Va Ahamiyyat-e Āb Dar Āsār-e Erfāni-e Fārsi [A review on the concept and importance of water in Persian mystical literature]. *Motāle'āt-e Naghd-e Adabi, 13*(2008), 47–80.
Spence, L. (1926). The gods of Peru. *The Open Court, 7*(1926), 398–411.
Streubert, S. H., & Carpenter, D. (2007). Qualitative research in nursing: advancing the humanistic imperative. Lippincott Williams & Wilkins.
Varner, G. R. (2009). *Sacred wells: a study in the history, meaning, and mythology of holy wells & waters*. New York: Algora Publishing.
Wright, D. P. (2009). Inventing god's law: how the covenant code of the Bible used and revised the laws of Hammurabi. Oxford University Press.
Zowghi, S. (2012). Zibāyi-shenāsi Dar Erfān-e Zartoshti Va Moshtarakāt-e Ān Bā Erfān-e Molānā Jalāl al-Din Balkhi [Aesthetics of zoroastrian mysticism and its similarity to Rumi's Mysticism]. *Tafsir Va Tahlil-e Motoon-e Zabān Va Adabiyāt-e Fārsi, 4*(13), 283–302.

Chapter 2
Production and Water Culture

Abstract This chapter examines the cultural values of water division systems in the Iranian local communities. Water division system has been evolved by the local communities over hundreds of years ago in order to better adapt to their harsh environment. This chapter shows how social dynamics and cultural potentials play a significant role in the water division system to ensure an efficient irrigation for every stakeholder. A village in western Iran has been selected as a case study to show how the qanat's water division system has remained almost intact, despite the profound changes that its management elements have undergone over the past few decades. In the traditional feudal system, the qanat water was divided among 11 agricultural units on an 11-day irrigation cycle that is still in use, though the feudal system itself is long gone. This chapter investigates water in relation to other production factors, which formed a specific feudalism in the village. In Iranian villages, feudalism was not identical in character, but it varied from region to region based on the relationship between the lords and their peasants. This relationship that underlay the entire agricultural system was contingent on five production factors: workforce, plough, seed, land and eventually water. This chapter tries to discover the role of water in the village's power relation. This chapter also shows how an irrigation technique like qanat could have served as a vehicle to transfer cultural elements from an ethnic group to others in the village. The village has been a historical stage on which different ethnicities have played their roles and then disappeared, but some vestiges of their cultures have remained in the shape of irrigation jargon. The qanat of the village is part of a bigger cultural landscape that goes back to the dawn of history.

2.1 Introduction

The village of Qasem-Abad belongs to Hamedan township, Hamedan province in western Iran. According to a census taken by Iran Statistics Center in 2006, the village's population amounts to 5661 with 1447 households. Qanat system is the most important water source in this village, which is a holdover from thousands of

M. Labbaf Khaneiki, *Cultural Dynamics of Water in Iranian Civilization*,
https://doi.org/10.1007/978-3-030-58900-4_2

years of sedentary livelihood in the region. Qanat consists of some shaft wells interconnected by a subterranean tunnel that drains out the groundwater seepage and conveys it onto the earth surface using the height difference between the two ends of the tunnel. Over the past centuries, various ethnicities dwelled in this area and left behind some vestiges of their culture. Hence, a wealth of cultural elements has been amassed over time, which is manifest in water and irrigation culture in this village.

Qasem-Abad has long been irrigated by two qanats that were the only water sources for both domestic and agricultural uses. The village has had no access to any surface streams like permanent rivers. Like the city of Hamedan itself, Qasem-Abad nestles on an alluvial fan in Zagros Mountains whose peaks receive a considerable annual precipitation. Airport meteorological station that is only 6 km from Qasem-Abad has recorded an annual precipitation of 206.3 mm on average (Hamedan Meteorological Organization 2017), whereas this number reaches between 400 and 500 mm (Hamidi and Hamidi 1997) mostly in the shape of snowfall (Ildermi 2012) at the top of the mountains overlooking the village. Therefore, such a disparity in precipitation types and amounts along with topographical and geological conditions has set the stage for qanat system in this region. In the elevated areas, the melting of the winter snows plays a vital role in replenishing groundwater reserves that have formed inside the alluvial fans and sediments of Alvand Mountains. In fact, the qanat of Qasem-Abad could have proved to be a solution to the region's uneven distribution of vital production factors like water and soil. At higher altitudes, water is abundant but arable soil is at a premium, and the climate does not lend itself to agriculture. On the contrary, at lower altitudes, water is in short supply but fertile soil is ample and the climate is well suited to agricultural activities. Hence, qanat can somehow even out such resources by bringing water from higher altitudes to the less elevated lands where suitable soil and climate allow a lucrative agriculture but in the presence of water.

The village of Qasem-Abad actually enjoys two qanats named Emamzadeh and Bozorg. Emamzadeh qanat with a length of 2 km appears on the surface close to a shrine, and its water is used for irrigating a vast area of orchards and ash trees. In 2017, when I first visited the village, this qanat was completely out of water due to years of negligence, collapse and tunnel obstruction. The water shareholders have turned to other water sources like tube wells in the area to irrigate their orchards. They have to rent some water shares from those tube wells as a last resort.

Although Emamzadeh qanat has been abandoned nowadays, the exit point of this qanat used to be one of the best places for recreation in the village in the past. A poem by Davar-e Hamedani bears witness to the fact that the qanat enjoyed a considerable discharge in the past, which drew many people even from the city of Hamedan for recreation.

Davar-e Hamedani was born in 1921 in Hamedan, whose first poetry was published in 1946 (Ebrahimi 2003, p. 35). I managed to find a few pages of a book entitled "Divān-e Kāshef al-Hayvah" written by Seyyed Hasan Shafayi Hamedani, where Davar-e Hamedani and his travel to Qasem-Abad get a mention as follows: "Davar-e Hamedani once went to the shrine of Qasem-Abad on a Friday for recreation. A large crowd of pilgrims flocked to the shades of the willow trees on both sides of a qanat water, such that he could not find an empty convenient place to sit down and rest. He binged on a delicious eggplant soup coupled with plenty of garlic and yoghurt, and then drank a lot of water, and remained sitting in the sun. He became sick in the sun, but he vented his rage on the villagers of Qasem-Abad and lampooned them in one of his poems" (Shafayi Hamedani).

Qasem-Abad's shrine is much revered, since it is a Shia saint' tomb, named Ghasim ibn Mohammad Bahtayi that descended from Imam Hossein (Gharaguzlu 2009, p. 141). The village has probably been named after that same shrine, and some historical records give credence to the fact that the village was known by the same name at least at the time of Mongolian invasion. For example, Abubakr Ravandi has mentioned this village in 1203 AD as follows: "The king of Kharazmshah arrived in Hamedan and was crowned on 1 July 1194, and then subdued all Iraqis and disarmed them. His army reduced all villages to rubble. He ordered to build a palace somewhere between Dazaj and Qasem-Abad, which was quickly accomplished only in one month" (Azkayi 2001, p. 291).

Building a sacred place like Qasem-Abad's shrine close to a qanat flow is rooted in the ancient culture of the region. Before Islam, it was quite common to build fire temples near a spring or qanat water as a sacred element that was highly venerated by Zoroastrians, and the same custom lived on later in the shape of Islamic shrines or mosques. For example, in Baneh, province of Kurdistan, Pir Morad tomb situated close to a natural spring has once been a fire temple that retained its religious function even after the inhabitants converted from Zoroastrianism to Islam (Mokhles et al. 2013, p. 24). Also, such shrines as Zobeydeh Khatoon in Naraq, Emamzadeh Ebrahim in Karaj, Chartaghi in the village of Nashlaj in Niyasar in Kashan, and Emamzadeh Hamzeh in Ab-Ali were all probably Sassanian fire temples that were converted to the Islamic sacred places and continued to be revered in a new ideological system (Javadi 2013, p. 46). Therefore, proximity of Qasem-Abad's shrine to the qanat exit point is expressive of an enduring influence that Iran's ancient culture exerts, regarding spiritual relationship between water and sacred places (Fig. 2.1).

2.2 Historical–Cultural Values of Qasem-Abad Qanat

The history of Qasem-Abad qanat and the history of Hamedan go hand-in-hand, and have always been inseparable. Many historical records highlight Hamedan as one of the oldest cities that Persians have ever founded. The fame of Hamedan is mostly indebted to Medes tribes who came together to form the Median Kingdom in the eighth century BC. Hamedan that was called Ecbatana at the time retained its importance as a summer residence under the Achaemenid Persian kings. Nevertheless, Hamedan's history goes back to as early as stone and copper ages, between 6000 and 3100 BC. During Bronze Age (3100–1500 BC), Hamedan enjoyed an urban development to which the archeological site of Pisa northwest of

Fig. 2.1 Emamzadeh qanat that has run out of water due to negligence and tunnel obstruction

the present-day Hamedan bears witness. The archeological excavations in Pisa revealed a variety of evidences related to the Bronze Age, and the upper layers were awash with Parthian objects. Although Hamedan precedes all known historical empires, the oldest written mention of Hamedan has been discovered in the inscription of Nabonidus who was the last king of the Babylonian Empire between 556 and 539 BC (Molla-zadeh and Taheri Dehkordi 2011, p. 6).

The next mention of Hamedan has been found in the Bistoon inscription where Darius the Great describes his victory in 520 BC suppressing a Median rebellion in Ecbatana as the rebels' headquarters. The Greek historian Herodotus (484–425 BC) provides us with valuable information about Hamedan and its development by Deioces as the Medes' capital. Polybius another Greek historian (208–125 BC) provides more detail about Hamedan that had risen to fame for horse breeding and grain production. Ctesias the Greek physician and historian who served at the court of Persian kings from 416 to 398 BC ascribes the foundation of Hamedan to the legendary Assyrian queen Semiramis as follows: "when Semiramis arrived in the region of Hamedan, on a flat plain she built a palace more gorgeous than any other palaces she had ever erected. She ordered that a long canal be dug from Alvand Mountain to Hamedan to supply water" (Ibid, p. 7). Ctesias's account shows that even in ancient times the prosperity of Hamedan plain was contingent on the water that had to be transferred from Alvand elevations. What Ctesias says about digging a long canal may be an allusion to the tradition of qanat construction that had long been common in the region. We cannot claim that Ctesias actually means qanat by mentioning such water transfer structures, until we look into the original text to see what word has been translated into "canal". What we can take for granted is the fact that qanats and their irrigation network have always played a crucial role in

Hamedan's agricultural production systems. For example, in Ecbatana the archeological excavations unearthed a network of water canals meandering along the streets of an urban fabric that dates back to the Parthian and Sassanian periods. The remains of those canals are expressive of the important status of water supply systems in the city's socio-economic dynamic (Hatefi Shoja' and Zareyi 2017, p. 34).

After the occupation of Muslim Arabs, Hamedan retained its political and economic status. The ruler of Hamedan could spare the city from the Arab invasion by signing a peace treaty with them in 644 AD, though the peace did not last long and a battle eventually broke out, in which Hamedan fell into the hands of Arabs. Although Hamedan remained as an economic hub, its population started to dwindle. Since mid-ninth century, Hasani people who were the descendants of Prophet Muhammad migrated to Hamedan in droves until they took control of the city. Their reign lasted for four centuries as the dynasty of Hamedan's Alawites. In 891 AD, Hamedan has been described as a huge city that afforded to pay an annual tax of 6 million Dirhams, which is a tremendous amount of money that was collected from the farmers and traders. From 931 AD onward, Hamedan was repeatedly invaded by Daylamites, Hasanawayhid and Kakuyids. The invaders demolished the city's walls and fortification to make its inhabitants surrender. At the time, there were two stone lion statues that were gorgeously built on the city's main gate, but both were smashed to pieces by the invaders. Many geographers and travelers who visited Hamedan in tenth century described it as a huge city with an area of 6 by 6 km, which enjoyed a city wall, fortification, watch towers, four iron gates, adobe houses, springs, water ditches and beautiful gardens (Zareyi 2011, p. 61).

The historical records leave no doubt that Hamedan has always enjoyed plenty of qanats and springs (Hatefi Shoja' and Zareyi 2017, p. 34) and its urban organization was under the influence of water resources and their geographical distribution. In Hamedan, over the course of history, geometrical formation of urban fabric has been affected by six factors: food supply, access to greenery, social bonds, security, livelihood and eventually water supply, especially qanats (Farshchian and Balali Oskooyi 2015, p. 64). "Garden" is one of the important elements of Hamedan's urban and rural fabrics, which has been developed in close connection with qanats. Florence Convention views "garden" as an inseparable element of the cultural landscape, whether it is located in the urban or rural areas (Hatefi Shoja' and Zareyi 2017, p. 32). In Hamedan, qanat could have changed the city's cultural landscape and created a different human ecology, which has defied the natural limitations.

Qanat hinges on the idea of transferring water from the rainy Alvand Mountains to the dry plain which has been home to many civilizations for thousands of years. Transferring water sets the stage for other urban elements like gardens that have always played a vital role in the region's socio-economic structures. For example, Nazari garden built in the early 1920s was irrigated by a qanat entering from its eastern side, though this garden is watered today by a tube well. The physical structure of the garden has remained unchanged except for its water source that has shifted from qanat to tube well (Hatefi Shoja' and Zareyi 2017, p. 39).

6.5% of Iran's qanats are located in the province of Hamedan, of which 227 qanats are identified and one has been registered as national heritage (Yekta et al. 2014, p. 3). The qanats are imbedded in the urban fabric of Hamedan in two ways: some qanats are running along the main axis of the city quarters, and some others are meandering through the smaller alleys across the city. In the periods of Safavids and Qajars, such important buildings as palace, mosque, caravansary, bazaar, public bath and dignitaries' houses were all built on the course of qanats for the purpose of easy access to water through some stairways named Pāyāb. Thus, the urban morphology was dictated by the trajectory of the qanats that meandered through Hamedan. The houses built on the course of qanats usually enjoyed a special structure named Sardāb that resembled a basement across whose floor a water ditch was running. The interaction between qanat and city was not unilateral. In other words, it was not only qanat that influenced the urban fabric, but qanat was in turn affected by the morphological and social traits of the area where it ran (Ibid, p. 6). For example, the rich had enough money to change the course of a qanat running near their houses, so that they could enjoy a private stairway to the qanat water. They had the qanat workers dig a bypass tunnel in order to bring its water as close to their kitchen as possible. Therefore, social status could affect the morphology of a particular qanat.

Qasem-Abad qanat is part of a bigger cultural landscape that goes back to the dawn of history. I could not track down any precise historical evidence helpful to clarify the antiquity of this qanat. All I could find during my field study was a piece of a stone structure that seemed to be a piece of a broken pulpit or a banister. The qanat workers used that piece as masonry in a reinforcing wall inside the qanat tunnel. The figures carved on this stone piece suggest that it dates back to the beginning of Timurid period around 1370 AD.[1] This stone can prove that the qanat-reinforcing wall is not older than Timurids, but it does not help to date the qanat yet.

Local data imply that the history of Qasem-Abad qanat even precedes the history of Turkic people in date, who gradually populated the region. I had the chance to interview Haj Gholam Torabi born in 1928. He was the only survivor of a generation of qanat chiefs who were in charge of the qanat's maintenance and rehabilitation. According to him, some 200 m upslope from the village communication center, three tunnels join and form the main qanat tunnel that runs toward the village. He is the last one who has seen the joining point and cleaned the tunnel from the village until there. Even he has not set foot in any of those three tunnels that bring water from different directions. He said that the people of Qasem-Abad knew only how to clean the qanat tunnel, but they had no idea of how to extend the tunnel into the aquifer or dig a new side tunnel to increase the qanat discharge. He quoted his father as saying "past people have dug whatever tunnels they needed, and we do not need to dig any more". Hence, a historical gap has yawned between

[1]According to Mr. Rajabali Labbaf Khaniki, Islamic period archeologist, and Dr. Meysam Labbaf Khaniki who specializes in Sassanian period archeology.

the existing qanat tradition in Qasem-Abad and what was once common in days of yore. A job-like tunnel extending is foreign to the present inhabitants of Qasem-Abad, and they do not remember anything akin to digging a new tunnel into the aquifer.

Most of the words referring to qanat works are Persian in origin, despite the fact that all the villagers are Turkish speakers. Most of the qanat jargon being used in Qasem-Abad is deeply rooted in the history of Persian language. For example, they use a measurement unit named Vargān that equals 60 L of water per second. In Qasem-Abad, it is very common to use Vargān to measure water flow in a ditch, when it comes to water division. Varg has originated in Pahlavi language or Middle Persian that was widely spoken in Iran before the fall of Sassanian Empire in the seventh century AD. Dehkhoda (1998) defines Varg as "a dyke that is built to curb floods. Also it is a barrier of wood, grass, soil and mud to divert a water flow". Rudaki (859–940 AD), the first poet of the Modern Persian language, has used the word Varg in his poetry as follows:

Āb Har Che Kamtarak Niroo Konad
Band-o Varg-e Sost-o Poodeh Befkanad
[Even a little weak current of water can destroy a flimsy dam and a frail Varg] (Rudaki 2017)

The word Vargān is made up of "Varg" and "Ān". Ān is a suffix that means land or place when it comes after a noun, like Iran that is composed of "Īr" or "Ēr" and "Ān", which means land of Aryans in Middle Persian (Davar 1953, p. viii). Hence, Vargān means a ditch where a small dyke or barrier is placed to divert water to another ditch. In Qasem-Abad this word is simply referred to as a water flow. Vargān is a remnant of an ancient irrigation jargon that was once common in the region, and is still living on in the language of those who do not speak Persian at all. But nowadays Persian speakers have put aside this word, and adopted such western words as Débit or Discharge to refer to a water flow. There are many of such Persian words that are still being used by the farmers of Qasem-Abad, while speaking in Turkish about qanat and irrigation. I cannot come up with a full glossary here, but a few examples can be helpful as follows:

Kariz: Kariz means qanat, which has been defined by Anjoman-Ara Persian dictionary as "an subterranean water conduit that is composed of two Persian words; Kah (wheat straw) and Riz (drop). The word kariz is derived from an irrigational job in which some straw is dropped in the water to measure its flow" (Hedayat 1871).

Ākhar Kolang: This word is referred to as a tunnel dead end where the qanat gallery terminates in an aquifer. This place should be extended into the aquifer if the farmers wish to extract more groundwater.

Komush: This word is another pronunciation of Kumesh that has been defined as "qanat digger" by all Persian dictionaries like Dehkhoda (1998), Moyin (1972), Borhan (Khalaf-e Tabrizi 1938), Anendraj (Mohammad Padeshah 1957) and Nafisi (1964). This word is still used in the same meaning in Qasem-Abad. It is said that the derivation of Qumes (city of Semnan) is from the word Kumesh due to a large number of qanat diggers who lived and worked in that city. Haghighat (1983) contends that people of Qumes had to dig many qanats, because their city was devoid of any surface stream or river. That was why they called their city Kumesh that was later mispronounced as Qumes.

Safar: This word literally means trip in Persian, but in Qasem-Abad it is referred to as irrigation cycle over which the water shareholders take turns irrigating their farmlands.

Juq: This word literally means ditch in Persian, but in Qasem-Abad it is referred to as a 24-h water share on an irrigation cycle.

Sar Juq: This word is composed of two Persian words: Sar meaning head and Juq means a 24-h water share, which together mean someone in charge of water division over a 24-h period.

Owgah: This word literally means "place of water" in Persian, but in Qasem-Abad it is referred to as sluice gate that can be opened or closed to control or divide water.

Peshk: This word means drawing lots to determine everyone's turn to take their water shares. In Persian dictionaries, Peshk has been defined as drawing lots in general.

Ravān keshi: This word means cleaning the qanat tunnel, which should be done once in a while to ease the water flow.

Dahaneh: This word literally means mouth in Persian, but in Qasem-Abad it is referred to as the exit point of qanat, where water appears on the surface.

Chābāneh: This word means qanat well, which is Persian in origin.

Māqār: This word is referred to as a hand-dug tunnel with no lining or masonry, but a tunnel shored up with stones and masonry is called Kureh. In Persian dictionaries, Māqār has been defined as cave and underground cavern.

Taleh push: This means a tunnel whose ceiling has been covered with heavy flat rocks.

Abundance of such Persian words in the language of Qasem-Abad's farmers may be expressive of a demographical change that has occurred in the region over the past centuries. Hamedan's fertile soil and favorable conditions have always drawn a variety of ethnic groups with different levels of political and cultural hegemony. The earliest story is about a pastoral community who migrated from Caucasus to the southern lands in the early third millennium BC, spreading across a vast area from eastern Anatolia to Urmia Lake basin. A group of that community continued to migrate southward along the Zagros Mountains to Hamedan, and then filled the vacuum in the region where the native population had already turned to agriculture and left the vast pasturelands unutilized. That community established a close commercial relationship with the region's other ethnic groups, a relationship that paved the way for them to assimilate into the native population and contribute to the formation of Zagros's Bronze Culture around 2400 BC (Motarjem 2011, p. 145). Apart from the aforementioned migration, the most important wave of migration to Hamedan region was associated with Turkic tribes who now make up 26.2% of Hamedan's urban population according to the 2013 census (Khatibi 2015,

Fig. 2.2 A piece of a broken pulpit or a banister found in a reinforcing wall inside the qanat tunnel

p. 145). After Turkmen tribes conquered the city of Rey and in the wake of a peace treaty between them and Abu Kalijar, the Buyid king, a large group of them migrated from Rey to Hamedan in 1039 AD. Buyid dynasty took advantage of Turkmen mercenaries to defeat their enemies, and that policy resulted in a bigger surge of Turkmen migrants who flooded Hamedan where they later assimilated into the local population (Momeni et al. 2013, p. 29). Migration of Turkic people who once enjoyed a pastoral economy and their assimilation into the agrarian local population caused their Turkish language to be saturated with the Persian words related to qanat and irrigation. Therefore, one may conclude that the qanat of Qasem-Abad is even older than the people who live in the village nowadays. Those words are in fact some units of information, which have been passed down from individual to individual and even from ethnicity to ethnicity over time. Those units of information are called meme (Aaen-Stockdale 2017; Wah and Shukri 2013), which seem to be the driving force behind the cultural evolution of Qasem-Abad's inhabitants along with their other memes (Fig. 2.2).

2.3 Water Ownership and Management System of Qasem-Abad Qanat

Given that water has always been one of the most important agricultural production factors, there has been a close correlation between the ownership of qanat as available water source and the conditions of agricultural production. Until the early 1960s, feudalism was the common production system in my study area. However, what I mean by feudalism structurally differs from western classic feudalism in which the vassals were obliged to give military services in return for the fief granted to them by the lord. This version of feudalism had a more productive nature that revolved around the economic relationship between a lord and the peasants who enjoyed agricultural tenure.

Even in Iran, feudalism was not identical in character across the country, but it varied from region to region based on the relationship between the lords and their peasants. This relationship that underlay the entire agricultural system was contingent on five production factors: workforce, plough, seed, land and eventually water. Workforce refers to the peasants who had agricultural tenure bestowed on them by a lord in the village. Some scholars view this kind of agricultural production system as exploitative, since it equates humans with other physical factors of production and treats them in a materialistic manner. Although in many cases this type of lord–peasant relationship proves to be oppressive, this relationship sits on a spectrum whose one extreme denotes extensive exploitation and full control by the lords, suffocating all economic and social aspects of the peasants. However, the other end of this spectrum is akin to a mutual agricultural contract in which the lords and the peasants just share their possibilities to reap a better harvest, without ending up in the lord's accumulation of power. All types of feudalism in Iran sit somewhere on this spectrum between the abovementioned extremes. If we call the full exploitative state "hard extreme" and full cooperative state "soft extreme", we can say that majority of qanat-based agricultural systems tended to sit closer to the soft extreme. This tendency was associated with both geographical conditions and socio-economic mechanism of qanat, which led to a lax lord–peasant relationship. Many of the qanat lords did not reside permanently in the village, but they were more involved in other economic activities like trading in the cities than agriculture in the villages. For them agriculture was only an offset economy in which they could invest their income surplus. Hence, their relationship with their peasants was more of a capitalist nature than feudalist.

I believe that the disparity between different degrees of lord–peasant relationship has much to do with the level of each party's control over the five production factors, including workforce, plough, seed, land and water. For example, the more production factors a lord had in his hands, the more authority he could exert over the peasants, and the closer the lord–peasant relationship moved toward the hard extreme, and vice versa. We can point out many examples from across Iran to substantiate this hypothesis, though we should try to find tenable explanation for some sporadic exceptions as well (Fig. 2.3).

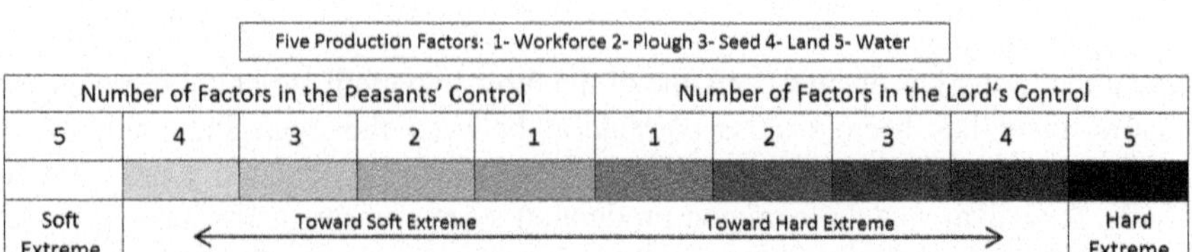

Fig. 2.3 Different sorts of agricultural feudal system on a spectrum of control over production factors

A hypothesis about the spectrum of agricultural feudal systems helps us better understand the situation of production systems and accordingly the ownership and management of Qasem-Abad qanat before the 1962 land reform. In the village of Qasem-Abad like many other villages whose agricultural production systems were dependent on qanat irrigation, the agricultural feudal system was inclined to the soft extreme that signified peasants' more control over production factors. In Qasem-Abad, the lord was entitled to one-fifth of the harvest (mostly wheat). In other words, when the farmers reaped their wheat harvest and finished winnowing, they took four-fifths of the crop as their wage and the rest went to the lord. In case of fruit gardens, the lord's share was one-third of the crop. The lord's share was in fact associated with the amount of his control over the production factors, because he could usually profit from the final revenue proportional to the number of production factors provided by him. For example, if a lord was in possession of water, land and seeds, it meant that three out of five production factors were provided by him, and then he could take three-fifths of the harvest in the end. In this case, the peasants got two-fifths of the crop, for they contributed by supplying workforce and plough.

In Qasem-Abad, the lord could get only one-fifth of the harvest, for he had control over one production factor. Nevertheless, we cannot make it clear what production factor was really in the lord's control, and how the lord contributed to the production process. It seems that the case of Qasem-Abad has been more complicated than the similar villages, when we try to place its agricultural feudal system on the aforementioned spectrum. In Qasem-Abad, the lord did not enjoy an absolute control over any of the five production factors, but he had a relative control over both water and land altogether. Although the lord was seemingly in possession of water and land, the peasants' active role in managing those factors could dilute the lord's absolute control. The lord's partial control over water and land led him to earn only one-fifth of the harvest, which was considered proportional to the total control he had over the production factors.

Before the 1962 land reform, in Qasem-Abad all cultivable lands were divided into 11 units locally named Juq. Each unit was headed by someone who was in charge of all agricultural affairs. Although those lands were formally in possession of the lord, the peasants were entitled to agricultural tenure or traditional right to cultivate on those lands, a special right that could be bought or sold among the peasants. All tenure-holders were organized and distributed across those 11 agricultural units. Each unit belonged to a group of farmers who worked together on the same land and then shared the crop. Traditionally, they had the right to sell or buy their tenures, while recognizing the lord's overall ownership. Hence, the lord partially shared his control over his lands with the peasants through tenure system that created a layered ownership. The lord owned the land as a physical boundary and the peasants owned the same land as a right to cultivate. What distinguished this tenure system from a lease contract was the permanent right that belonged to the peasants, which could be inherited, sold or bought independently of the lord's ownership.

Thus, this kind of joint control over the lands did not set the stage for the advent of a lord who could tether the peasants to the lands as we have seen in many parts of Iran. Before the 1962 land reform, Gholam Hossein Qareguzlu was the last lord 32 ha of whose lands were re-distributed among the tenure-holders, and 230 ha were purchased and changed into a garden named Farah Pahlavi by the then government. In addition, 100 ha of the lord's lands were allocated to an airport project that was under way by the government. It is worth noting that apart from the tenure-holders who permanently worked in the 11 agricultural units, there were more people involved in agricultural activities in the same units as daily-wage laborers called Āftāb Neshin (basking in the sun) who were occasionally hired by the tenure-holders. Qanat diggers fell into this category of workers; Āftāb Neshin.

The same situation applied to the ownership and management of water, since water and land ownerships were inseparable in the village of Qasem-Abad. The duration of irrigation cycle correlated with the number of agricultural units. As mentioned, the qanat shareholders irrigated their farmlands by turns over a certain period of time that was called irrigation cycle. In Qasem-Abad, every irrigation cycle was 11 days, and each shareholder's turn came once every 11 days. Each day was allotted to an agricultural unit, when all tenure-holders teamed up to irrigate every corner of their land. The same relationship regarding land between the lord and his peasants was in place in terms of water. The qanat was maintained jointly by the lord and the peasants, and the control over its water was in turn shared among them. The qanat of Qasem-Abad needed to be cleaned out once a year in order to sustain its function. The cleaning of qanat was done by a team of four workers of whom one was a qanat master called Komush and the others were unskilled laborers. The qanat master was hired and paid by the village lord through a contractor who specialized in finding appropriate qanat masters and overseeing their works. The contractor was called Sar-Komush who undertook the qanat master's wage by the money that the lord paid him for brokering and supervising the qanat project. One of the well-known qanat contractors has been Osta Asadollah who was repeatedly mentioned in the interviews I had with the villagers in 2019. Although the village of Qasem-Abad was home to such good qanat masters as Gholam Ali or Esmaeel, the quarter of Joulan in Hamedan was more famous for having adept masters.

For cleaning the qanat, which was carried out between October and March every year, three teams worked shifts to speed up the project. Between October and March, the qanat discharge was at its lowest level, which made it easier for the workers to remove the mud and dirt from the qanat gallery. Moreover, the cold seasons between October and March brought cultivation to a halt, when the farmers had more spare time to handle the qanat job. The lord undertook only one worker as qanat master on each team, and the other workers had to be supplied by the tenure-holders as their contribution. Hence, each agricultural unit was obliged to send three workers for each team, and when their shift was over, they were replaced by three workers from the next unit. This way, all units contributed to the cleaning of qanat by turns over the abovementioned period. If a tenure-holder was not able to fulfill his commitment by working in the qanat for any reason, he paid someone else to work on his behalf.

It is worth noting that the tenure-holders' cooperation was not confined only to the cleaning of qanat, but the same cooperation was deployed in repairing and cleaning the ditches and irrigation canals. This cooperation is still extant, which is locally called "khāna passā". Every year in late April, all tenure-holders came together to determine the amount of contribution that each person should have made to the cleaning of ditches, based on each person's land/water share. Those who were not physically fit or could not participate for any reason had to send someone else on their behalf. The tenure-holders' contribution could give them a better status, when it came to having control over water as a production factor.

In 1947, a worker that dragged the buckets of mud out of the qanat tunnel received a wage of 50 cents per day. In the same year, a qanat master was paid 3 Tomans equivalent to 1.5 US dollars per day. All those activities were done in the light of a systematic relationship between the lord and the peasants in Qasem-Abad. However, in the wake of the 1962 land reform, the functional relationship between the lord and the peasants was disturbed and a series of drastic changes took place in the agricultural production system, which in turn affected the qanat of Qasem-Abad. In many references related to the land reform, re-distribution of lands among the tenure-holders and the collapse of the agricultural feudal system were discussed as one of the important reasons behind the annihilation of qanats in Iran (Hojjat 2006, p. 78). The land reform had the same impact on Qasem-Abad, and the village's qanats bore the brunt of its consequences.

After Gholam Hossein Qareguzlu handed over his lands according to the 1962 land reform law, his waning traditional role as the village lord caused disturbance to the qanat management system. As mentioned, before the land reform, the cleaning of qanat was done periodically and on a regular basis under the lord's supervision. Every year the lord took the lead in cleaning the qanat by hiring the required workers and providing such tools as windlass, pickaxe, bucket and shovel, and then the tenure-holders offered their contribution by sending three workers for each team.

The lord oversaw all stages of the qanat project, since any mistake could hamper the qanat that was their main source of water. When the lord was removed from the production system, the tenure-holders suddenly lost the axis around which they were systematically mobilized and organized. The tenure-holders developed more divergence and individuality, which was against the cooperative soul of qanat. Thus, the land reform and the lord's absence brought the qanat project to a standstill, such that the cleaning of qanat was suspended for about 15 years and the qanat workers lost their job thereafter. In 1976, the battered management system of the qanat regained its balance but in a new composition with new functional relationships. The tenure-holders who now became petty landlords tried to take the place of the gone lord, and took the helm of the qanat rehabilitation project. Years of negligence and poor maintenance had led to a dramatic decline in the qanat discharge, and the qanat's new owners had to put their heads together to find a solution. According to the elderly locals, before the land reform, the qanat water was so much that it could not only satisfy their village, but could also irrigate all lands of a neighboring village named Jourghan just over 24 h. When Jourghan ran short of water, they fastened their hope on the qanat of Qasem-Abad whose water was ample.

Hence, they selected someone named Haj Gholam Torabi as qanat custodian. He was responsible for all qanat's maintenance affairs, by hiring the qanat workers, supervising them and handling the finance. He calculated how much each owner had to contribute to the qanat maintenance project, proportional to the amount of their water shares. The collected money was spent on the cleaning of qanat. Unlike the lord's time, the qanat owners no longer worked in person in the qanat project, and they preferred to contribute by money. The new socio-economic transition that had broken down a big lord into several smaller lords made them feel that it was beneath them to do such menial work as qanat cleaning. Thus, a new system for qanat maintenance came into being and evolved.

In the new system, the cost of maintenance was divided among the qanat shareholders and was collected from them. Although the new system resurrected the qanat of Qasem-Abad, another problem remained to be solved: deficiency in qanat workers. 15 years of the qanat cleaning shutdown frustrated the qanat workers, many of whom migrated to other regions in search of job. Their migration caused a gap in the chain of qanat knowledge that was traditionally passed down from generation to generation. Gholam Torabi, the new qanat custodian, had to sort out all those problems.

Since 1976, the cleaning of qanat was resumed and the qanat discharge was improved to a large extent despite the abovementioned problems. However, in 1997 another governmental program intervened and unbalanced the qanat management system for a second time. At the time, the government intervention was expanded to almost every aspect of rural life through establishing the village council. The qanat maintenance was put on the agenda of the village council that in fact replaced the traditional qanat custodian who raised funds and managed the whole process of the qanat cleaning. Therefore, the balance that the qanat managed to regain after the land reform was disturbed again by another top-down program. The village council was not very knowledgeable about the technical and social aspects of qanat, and they caused disturbance to the sustainable relationship between the qanat and its beneficiaries by injecting the oil money that came from government. The qanat of Qasem-Abad bears witness to the fact that government intervention in indigenous water management systems not only helps to improve their present situation, but it bogs down the socio-economic functions that such a system already enjoys in a traditional context.

In a nutshell, I view the water management of Qasem-Abad qanat as a historical process that has begun with an agricultural feudal system in which the peasants' contribution has played a pivotal role under the lord's supervision. The qanat management system proved to be efficient owing to the fact that the qanat proper maintenance was to the benefit of both the lord and his peasants. The 1962 land reform ushered in a new system in which the lord was divested of his traditional role without envisioning any feasible alternative. Hence, the qanat sank into oblivion that lasted about 15 years. Later the qanat beneficiaries regained the balance between the new socio-economic condition and the qanat, and developed a new water management system in which responsibility was divided among the owners based on the amount of their water shares, unlike the former system in which the onus was only on the lord to supervise the qanat maintenance. The qanat owners had to pay an annual maintenance fee that was proportional to each one's water share. If someone rented a water share for a period of time, he was charged for the maintenance fee, and then the money was subtracted from the total rent.

Later another government intervention disturbed the spontaneous balance once again by introducing a rural council to the village. A managerial crisis along with a deluge of developmental projects that were hastily carried out without doing any feasibility or assessment studies took the qanat of Qasem-Abad to the brink of annihilation. Eventually, the qanat totally dried up in 2002. The qanat owners decided to request governmental aid, when they could not handle the problem of water scarcity and the ensuing income deficiency. In 2006, they opened a bank account in the name of their qanat and made a deposit, in the hope that they could get a loan for the qanat rehabilitation, but their loan request was turned down and the qanat remained deserted. Eventually, the interior ministry established a rural bureau that financed the qanat rehabilitation in 2014. That fund made it possible for the villagers to remove the obstructions from the qanat's water transport section, which resulted in water flowing down the dry tunnel after many years. Qasem-Abad's rural bureau is in the process of refurbishing more parts of the qanat, and again a different qanat management system is taking shape. In the village, the bureau established a farming council that legally aims to:

- settle disputes over farm boundaries and land transactions
- settle disputes over water and land ownerships
- keep records of water shares and water transactions
- handle the qanat maintenance and rehabilitation projects.

Hence, in this new management system, the qanat maintenance comes within the purview of a governmental bureau, which shows least resemblance to the traditional model. In 2016, the farming council was composed of five members named Heydar Alavi, Jafar Safari, Abdolhossein Ebrahimpoor, Nosratollah Talea' Hemmat, and Ali Askari. Once again the qanat of Qasem-Abad has returned to the production cycle by irrigating 350 ha of orchards. The historical process of the qanat management has been depicted in the following model (Fig. 2.4).

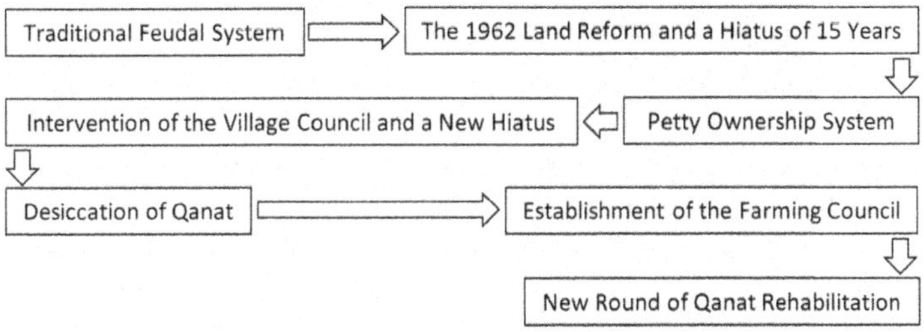

Fig. 2.4 History of the qanat management system in Qasem-Abad

2.4 Water Division System of Qasem-Abad Qanat

The qanat's water division system has remained almost intact despite the profound changes that its management structure has undergone over the past decades. In the traditional feudal system, the qanat water was divided among 11 agricultural units on an 11-day irrigation cycle. 24 h of irrigation was called a Juq, and 1 h of irrigation was called Nasaq. A Nasaq was in fact a certain area of land (3600 m^2) that could be irrigated by the qanat flow in 1 h. This terminology is reminiscent of the feudal system but is still valid in the village though much time has elapsed since that time. Unchangeable nature of the village's water division is associated with the farmers' traditional status in the agricultural production system. In this village, even at the time of feudal system, the farmers had effective control of the production factors, which gave them a better status compared to other villages where the lord had the last word. Hence, the qanat water was not in the lord's absolute possession, but the tenure-holders were entitled to its utilization. Qanat water was divided among the tenure-holders in a particular order. Given that the qanat utilization system remained unchanged after the collapse of the traditional feudal system, the qanat water is still divided based on the same traditional rules among the farmers who have now become petty owners (Fig. 2.5).

As mentioned, a period of 24 h for irrigation is called a Juq, a period of 12 h is called Yārem juq, which means half Juq, and a period of 6 h is called Roppe juq. These units make sense only in the framework of an irrigation cycle that is 11 days long in Qasem-Abad. On the irrigation cycle, each shareholder's turn is well determined, and when he receives his water share, he should wait for another 11-day period to irrigate his land for the next round. It is worth noting that the position of every shareholder on each day of the irrigation cycle is fixed. In other words, they cannot easily move their water shares from one day to another. However, the position of every day on the irrigation cycle is flexible and may vary from year to year. For example, if someone has a water share on the second day of the 11-day irrigation cycle, he is not allowed to forgo his turn on the second day and

Fig. 2.5 Interview with the qanat owners who have congregated in the village mosque

take it instead on the third day. However, next year, his water share along with that of his fellow farmers may be shifted from the second day to another day on the irrigation cycle. Every day is like an irrigation band whose collective position on the irrigation cycle changes from years to year. Every year all water shareholders congregate in the village mosque, and they draw lots to decide a new order in which the days should be re-arranged on the irrigation cycle. For example, in the year 2017, Mohammad Khavari, Mohammad Ghadiri, Ali Askari, Hossein Modari, Akbar Meshkingar, Hossein Besharati, Yusef Alibeygi and Gholam Ghasemi all got their water shares on the first day. All these persons could irrigate their lands just on the first day, and they were not allowed to get their water shares on the other days over the irrigation cycle, except in case of emergency when a farmer could temporarily swap water shares with another farmer on other days. If a person owns more water shares on different days of the irrigation cycle, he should use them separately in due time.

On a particular day (Juq), the Juq members took turns irrigating in a particular order. For example, on the first day, Mohammad Khavari irrigated first, then Mohammad Ghadiri, then Ali Askari, and so on. In the next round, they changed this order, so that everyone enjoyed equal condition, because some turns came by night and some by day. They could do justice by changing the shareholders' order in each round of irrigation. Table 2.1 shows how the order of shareholders is changed to treat everyone as fairly as possible. The first person in the round 1 goes to the back of the queue in the round 2, and the second person moves forward and so on.

The order of all 11 days is specified over the irrigation cycle by drawing lots at the beginning of the farming year in the late May if the qanat discharge is high, but in the late April if the qanat discharge is low. To do so, they write the names of the days' heads on 11 pieces of paper, and then a kid draws the papers one by one and reads out the names. Hence, they decide which day comes first, second and so on. The last day is locally called Dānqā.

Every Juq (day) has a head locally called Juq-bāshe who is responsible for fair water division among all members. He takes care of all irrigation turns and their changes in every round of irrigation. The head is selected from among those who have most water shares on the same day. Table 2.2 shows the order of all 11 days over the irrigation cycle in 2017.

Table 2.1 Changing turns on a day of the irrigation cycle over four successive rounds of irrigation

Turns / Rounds	1	2	3	4
1	Person A	Person B	Person C	Person D
2	Person B	Person C	Person D	Person A
3	Person C	Person D	Person A	Person B
4	Person D	Person A	Person B	Person C

Table 2.2 Division of water on an 11-day irrigation cycle in Qasem-Abad

Day number	1
Day head	Mohammad Ghadiri
Shareholders	Water share (hour)
Mohammad Ghadiri	6
Masoom Khavari	6
Ali Askari, Hossein Moradi and Akbar Meshkingar	6
Hossein Besharati, Yusef Alibeygi and Gholam Ghasemi	6
Total	24
Day number	2
Day head	Jafar Safari
Shareholders	Water share (hour)
Jafar Safari	6
Mohammad Faraji	6
Mohammad Yusefi	4
Seyyed Ali Salehi	4
Mohammad Hossein Khalili	4
Total	24
Day number	3
Day head	Mohammad Ali Ahmadi Khorram
Shareholders	Water share (hour)
Mohammad Ali Ahmadi Khorram	6
Aziz Fattahi	3
Morteza Tavakkoli	3
Safar Fattahi	4
Ahmad Mirzayi	2
Ali Farhadi Khorram	4
Ali Yarmoradi	2
Total	24
Day number	4
Day head	Nasrollah Talea' Hemmat
Shareholders	Water share (hour)
Nasrollah Talea' Hemmat, Gholam Khanabad and Mohsen Cherami	12
Asghar Karimi	4
Aziz Zareyi	4
Mirza Sheykhian	4
Total	24
Day number	5
Day head	Jafar Safari
Shareholders	Water share (hour)

(continued)

Table 2.2 (continued)

Jafar Safari	6
Nasrollah Talea' Hemmat	3
Ali Akbari	3
Khodayar Yari	6
Gholam Hossein Qareguzlu's heir	6
Total	24
Day number	6
Day head	Heydar Alavi
Shareholders	Water share (hour)
Heydar Alavi	5
Pasha Alavi	7
Heshmat Talea' Mohegh	6
Hasan Salehi	2
Mohammad Faraji	2
Gholam Rezayi	2
Total	24
Day number	7
Day head	Asghar Jafari
Shareholders	Water share (hour)
Asghar Jafari	6
Morteza Tavakkoli	4
Ezatollah Zavvari	4
Ghorban Ali Feyzi	7
Mohammad Asadi	2
Mohammad Mirzayi Kia	1
Total	24
Day number	8
Day head	Gholam Hossein Sheykh Mohammadi
Shareholders	Water share (hour)
Gholam Hossein Sheykh Mohammadi	4
Naser Shirian	4
Hasan Naderi	2
Ramezan Naderi	2
Soltan Ali Karimi	2
Ali Taheri	2
Gholam Reza Khatami	3
Mohammad Taheri	2
Hossein Sangari	3
Total	24
Day number	9
Day head	Hossein Karimi
Shareholders	Water share (hour)

(continued)

Table 2.2 (continued)

Hossein Karimi	4
Abbas Kooroshi	4
Gholam Ghasemi	4
Bahram Mosayyebi	2
Javad Rahimi	3
Akbar Sohrabi	2
Alireza Khatami	3
Abdol-hossein Ebrahimpoor	2
Total	24
Day number	10
Day head	Gholam Torab Ali (Torabi)
Shareholders	Water share (hour)
Gholam Torab Ali (Torabi)	5
Hossein Naderi	4
Eskandar Askari	4
Abdol-hossein Ebrahimpoor	4
Safar Fattahi	7
Total	24
Day number	9 (Dānqā)
Day head	Hossein Jafari
Shareholders	Water share (hour)
Hossein Jafari	3
Gholam Jafari	4
Karam Gohari	5
Azizollah Azizi	4
Ghorban Ali Feyzi	6
Abbas Sardari	2
Total	24

2.5 Water and Time

In the past in Qasem-Abad, they used a type of clepsydra or water clock named Peymāneh to calculate the time of irrigation. This tool was the reverse of a traditional water clock that was once used in the Iranian central regions like Yazd. In Yazd, a clepsydra consists of two bowls made of copper, one of which is so small that it can freely float on the surface of water in the larger one. The floating bowl has a tiny hole at its bottom through which water can enter the bowl and gradually fill it up. After being filled which may take a certain time, the small bowl sinks into the water and bumps at the bottom of the large bowl. As soon as the bump is heard, a unit of time is over, so the time between the two bumps equals a certain unit of time (Labbaf Khaneiki 2019, p. 50). However, in Qasem-Abad, a clepsydra was a pottery bowl at whose bottom there was a small hole. The bowl was placed on a wooden tripod or simply on three stones, such that water could drip down out of the hole. It took about 2 h until the bowl got completely empty of water. If they needed to calculate a time shorter than 2 h, they used a small stick the size of a pencil, named Laleh. They carved some marks on the stick and held it vertically in the bowl to gauge its water level that indicated the remaining time. For example, if the water level reached the middle mark on the stick, they realized that 1 h has passed so far. In Nehbandan, South Khurasan Province, the same clepsydra was called Khabiyeh. What distinguished their clepsydra with that of Qasem-Abad was that they did not use a pottery bowl but a copper bowl inside which eight parallel lines were cut to mark more time subdivisions (Labbaf Khaneiki 2006, p. 97).

Traditional clepsydra or water clock has different names in different parts of Iran, where irrigation water has to be divided by time. Pengān and Jām are among the most common names (Noorbakhsh 1987, p. 395). One can trace this kind of water clock back to the Manichaean tradition.[2] In 1983, Hassouri wrote an article on a Manichaean poem in which he came across the Middle Persian term "Jām ī Bōzishn", translated as "the time of freedom". In the poem, Mani has been metaphorically called the time or the moment of freedom (Hassouri 2014, p. 89). Jām as a name for water clock was abstracted from its original tangible usage and was applied for describing a concept as ambiguous as time. Hence, in the Manichaean literature, Jām refers to both time and its measuring instrument.

Nasir Khusraw (1004–1088 AD), a Persian poet, philosopher and traveler, uses the word Pengān in one of his poems where he expresses his astonishment at the idea that human can unravel many of the world's riddles by his growing wisdom. He is impressed by the human wisdom that has made it possible to calculate the eternity of time by means of a little holed cup (Pengān), as he says:

> In the past, who did ever know that one will be able to measure time just by a Pengān?
>
> Who did ever know that Moon, Jupiter and Saturn get their illumination from Sun light?
>
> Who did ever know that all seas, mountains and deserts are suspended in space without anything to lean on?[3] (Nasir Khusraw 1978, p. 83)

In an agrarian society whose economy has always been anchored in irrigated agriculture, it is not far beyond expectation that irrigation initially made the idea of time germinate in the farmers' minds. Water was their most important production factor, which was always in short supply in such an arid environment. Hence, their available water source should have been rationed as fairly as possible. Water division based on the flow volume was not a good idea, since a fickle climate and

---

[2]Mani (216–274/277 AD) was an Iranian prophet and the founder of Manichaeism that was strongly inspired and influenced by Gnosticism.

[3]Ke dānest az avval, che gooyi, ke yidoon
Zamān ra bepeymood shāyad be pengān?
Ke dānest kaz noor-e khorshid girad
Hami rowshani, māh o berjis o keyvān?
Ke dānest kandar havā bi-sotooni
Setādast daryā o kooh o biyābāan?

accordingly a fluctuating water source was another problem for them to tackle. If they divided water based on its flow volume, each shareholder became entitled to a certain amount of water, regardless of the total flow. In this case, when the total flow dwindled in the wake of a drought, it was very likely that each shareholder still expected to receive the same amount of water as he usually did as his certain water share. If all shareholders had the same mentality, conflict over water was inevitable, because the sum of all water shares exceeded the total flow available in the time of drought. To solve this problem, they preferred to divide water based on time, regardless of the total flow in the main canal. Thus, proportional to their water shares, all shareholders equally enjoyed more water if they were fortunate enough to have a bigger total flow, and equally received less water if the total flow dwindled. Irrigation actually brought the issue of time to the fore, and I believe that water division was the first point in the Iranian history, where they came across the concept of time in its entirety. That was why the Manichaean poet uttered an irrigation technical word to describe "time" in a context that actually had nothing to do with irrigation. Although the concept of time in the first place emanated from irrigation and water division, it later overwhelmed many scholars and scientists in Iran and Mesopotamia over the course of history, some of whom like Banu Musa brothers (ninth century AD) excelled at water engineering (Beyg Babapoor 2010, p. 396). Banu Musa's prominent student, Sabit Ibn Horrani (836–901 AD) wrote a treatise about timing and clocks, though Abu Jafar Mohammad Ibn Musa Kharazmi (800–847 AD) preceded him by writing a detailed book about sundial and time calculation (Azkayi 1999, p. 173). From twelfth century, Ibn al-Jazari (1974) and Abu Hatam Esfazari (Abattouy 2016, p. 17) are famous for their efforts in designing and describing different types of time measurement equipment. In 1324 AD, another scholar and engineer named Hossein Ibn Abubakr Amoli put up a novel huge water clock in the middle of Rokniyeh school in the city of Yazd (Noorbakhsh 1987, p. 396) (Fig. 2.6).

In this sense, water and time have always been interwoven in the Iranian highlands including the village of Qasem-Abad. In Qasem-Abad, the farmers needed to calibrate their water clock once in a while, because any little residue settled around the hole could hamper its work. In the past, they used a sundial to set their water clock. An accurate water clock was expected to get empty of water just six times between sunrise and sunset; otherwise the hole was modified wider or narrower to fix the water clock. It was a time-consuming and painstaking job to set a water clock that way. Hence, someone named Sheykh Reza for the first time brought a modern watch to the village. He died in 1946, but the villagers used to synchronize the water clock with Sheykh Reza's watch for a long time, the way that the bowl got empty only one time during 2 h. Later, modern watches came into vogue and made inroads into the traditional water clock, though the water division system remained intact.

In Qasem-Abad, two kinds of measurement units are used in their traditional irrigation system: volume unit and time unit. The volume unit is called Vargān, used to measure water flow in the irrigation network in order to divide water among different canals. A Vargān equals 60 L per second. After water comes out of qanat,

Fig. 2.6 A sophisticated water clock designed by Ibn al-Razzaz al-Jazari in twelfth century AD (al-Jazari 1974, p. 216)

it should be divided among several ditches each of which goes to a different area. The place where water is divided is called Owgah. Water is divided by a sluice gate that blocks water and diverts it to other ditches whenever needed. The sluice gate is an iron frame into which an iron rectangular plate is driven down to block the flow. The time unit is used to distribute water among the shareholders on an 11-day irrigation cycle. This unit measures only the time that each farmer has to irrigate their lands, regardless of how much water flows down the ditch. For example, if someone has a water share of 1 h, it means that he can divert water from the main ditch to his own land for 1 h (Fig. 2.7).

Unlike many regions where qanat is the main water source, in Qasem-Abad water and land ownerships are inseparable. Needless to say, water is divided only among those who are legally entitled to a certain water share. Hence, water ownership pertains to someone's right to use the qanat water for a certain length of time and on a certain irrigation cycle. In Iran, there are two types of water ownership in general; land-based water ownership and landless water ownership. Land-based water ownership is mostly applicable to rivers and natural springs. In such regions, water cannot be sold and bought independently of land, and water and land go hand-in-hand on any transaction. Hence, water division unit is based on the area of the irrigated land. In other words, water is measured by the area that it can potentially irrigate.

Nevertheless, in the qanats owned by a single lord, water and land ownerships are usually together. In such cases, qanat water is divided among several agricultural units based on their areas. Qasem-Abad qanat fell into this category since before the 1962 land reform, and it has retained its water–land ownership system to a large extent. Now water is not salable separately from land under any circumstances.

Fig. 2.7 Owgah where water is divided based on its volume

The area of cultivated lands closely correlates with the amount of qanat water. The more the qanat water, the more lands would be cultivated. It is easier to manage water affairs in a system of land-based water ownership. But the farmers have to devise a complicated water management, when it comes to a system of landless water ownership which is common in many parts of the Iranian central plateau. Landless water ownership is usually practiced for the qanats whose water is owned by a large number of petty shareholders. The right to the qanat water is divided into many small shares that are valued independently of the cultivated lands. In the local market, these shares can easily be sold or bought, as capital investment for any agricultural activity. Some people may buy water shares without having any land at all. Hence, this reliable market attracts some capital from outside the village. For example, some urban merchants who seek a reliable property to invest part of their capital turn to such qanats.

In the qanats with such a petty ownership, water right is divided based on time, by minutes and even seconds. Therefore, water division is a pivotal task that entails an intricate management system headed by someone named Mirāb. That is why such qanats are called Mirābi (Mirāb-related) qanats too. In such qanats, we should not expect a correlation between the owners' water shares and the area of lands in their possession. Someone's water share may be much more than what he needs to irrigate his lands, and someone's vast land may be disproportional to the small water share that he owns. This situation has something to do with their special water division system. Landless water ownership brings about different consequences including development of a pre-modern capitalism and failure in land consolidation programs. Water and land are not of the same economic value, and their prices fluctuate in different directions under different circumstances. For example, during a drought water price may go up and surpass land price, and during a wet year the reverse is true. Therefore, it would be very difficult to determine everyone's share of the total revenue, if the fragmented land parcels and the small water shares would be consolidated in order to increase the production efficiency. A person may have more water and less land, and he may claim more share of the total revenue especially when water is in short supply. However, it seems absurd that some people who contribute less land or even no land to a land consolidation program can benefit more than do those who are the real owners of the lands. In fact, it is very difficult to translate water and land values into each other under the changing circumstances to which they react differently. Moreover, those whose water does not suffice to irrigate their lands have to buy or rent water right from those who have hoarded more qanat water shares. Hence, a group of qanat shareholders can generate an income just by renting out their water shares without being involved in the local production system. However, this problem does not apply to the qanat of Qasem-Abad, because of its special traits.

In a nutshell we can say that water division based on time usually paves the way for landless water ownership, which makes it possible for the qanat shareholders to buy or sell their water shares separately from their lands. Nevertheless, in the village of Qasem-Abad, water and land ownerships are inseparable despite the fact that their water division system is based on time. In this village, water and land ownerships go hand in hand, and a land without water share is considered economically valueless. In the villages where landless water ownership is in place, the time unit for water division is broken down into as small subdivisions as minutes and seconds. However, the smallest time subdivision is 1 h in Qasem-Abad.

In the past, sundial was also used to calculate the irrigation time in Qasem-Abad, in the case of water shares longer than 4 h. Sundial was used to calculate the time of other agricultural activities as well, especially when they wanted to work shifts. In this case, they simply used their own fingers as sundial to keep track of time. The farmer held out his hand toward sun with all fingers inclined to his palm, and he raised his middle finger a little more at a certain angle to see its shadow on his palm. In the early morning, its shadow fell at the base of his thumb, but it moved around his palm and fell at the base of his middle finger at noon (Fig. 2.8).

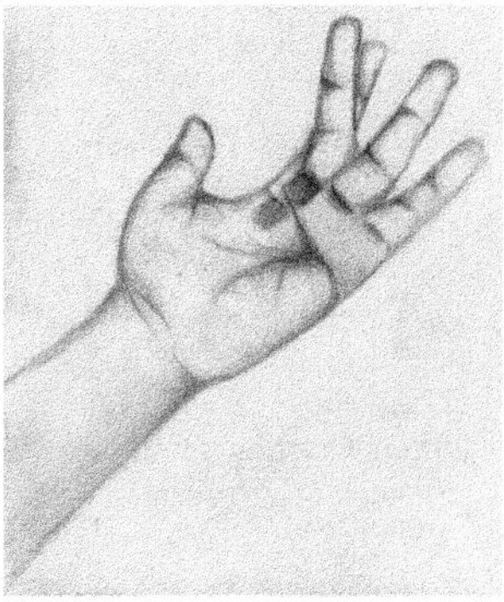

Fig. 2.8 Qasem-Abad's farmers used their hands as sundial to calculate the irrigation time (by Kiana Labbaf Khaniki)

2.6 Conclusion

In the Iranian highlands, water has always been the most crucial production factor. Water not only determined all socio-economic interactions within their production system but also contributed to their cultural evolution. In the eleventh century AD, Turkic tribes began migrating to Hamedan region where Qasem-Abad village was located. They came from central Asian steppes where their pastoral economy was dependent on pastures, to an agrarian territory where water had the final word. They had to get off their horses and adopt a new agricultural lifestyle that revolved around water and irrigation. Although they still spoke Turkish, they learned a Persian irrigation jargon from the native farmers. The native farmers were eventually replaced by the Turkish newcomers, though their irrigation memes lived on into the present time. The irrigation jargon turned out to be like cultural genes that transferred not only technical water knowledge but also a parcel of beliefs and customs somehow associated with water. Later on, water became the backbone of their cultural and social evolution in the village of Qasem-Abad. In the village, power relation hinged on water ownership, and water was divided along the borders between social classes.

References

Aaen-Stockdale, C. (2017). Selfish memes: an update of Richard Dawkins' bibliometric analysis of key papers in sociobiology. Publications, 5(12), https://doi.org/10.3390/publications5020012.

Abattouy, M. (2016). Resāle-hāye Abu Hātam Esfazāri Dar Elm-e Mekānik [Abu Hatam Esfazari's treatises on mechanical sciences]. (H. R. Nafisi, Trans.). Mirās-e Elmi-ye Eslām Va Irān, 5(1), 6–34.

al-Jazari, I.-R. (1974). *The book of knowledge of ingenious mechanical devices* (D. R. Hill, Trans.). Boston: D. Reidel Publishing Company.

Azkayi, P. (1999). Elm al-Hiyal Va Fonoon-e Ān [Science of tricks and its techniques]. *Tahghighāt-e Eslāmi, 1&2*(1999), 167–197.

Azkayi, P. (2001). *Hamedān Nāmeh [Book of Hamedan]*. Hamedān: Mādestān.

Beyg Babapoor, Y. (2010). Moarefi-ye Noskhe-yi Nafis Dar Mohandesi-ye Mekānik Az Sade-ye Sevvom-e Hejri, Neveshte-ye Banu Musā [An introduction to an invaluable treatise on mechanical engineering from the 9th century AD, by Banu Musa]. *Payām-e Bahārestān, 2*(8), 395–427.

Davar, F. C. (1953). *Iran and its culture*. India, Bombay: New Book Co., Limited.

Dehkhoda, A. A. (1998). *Loghat-nāmeh [Persian dictionary]*. Tehran: Moasseseh-ye Enteshārāt Va Chāp-e Dāneshgāh-e Tehrān.

Ebrahimi, P. (2003). *Simā-ye Mirās-e Farhangi-ye Hamedān [Image of Hamedan's cultural heritages]*. Tehran: Sāzmān-e Mirās-e Farhangi-ye Keshvar.

Farshchian, A. H., & Balali Oskooyi, A. (2015). Nemood-e Hendeseh-ye Edrāki Bar Seyr-e Takāmol-e Shahr-e Irāni Dar Advār-e Mokhtalef-e Shahr-sāzi: Shahr-e Hamedān [Manifestation of intuitive geometry in the evolution of the Iranian cities in the different periods of urbanization—case study: Hamedan]. *Motālea'āt-e Shahri, 15*(1394), 53–67.

Gharaguzlu, G. H. (2009). *Hegmatāneh Tā Hamedān [From Hegmataneh to Hamedan]*. Tehran: Eghbāl.

Haghighat, A. (1983). *Tārikh-e Qumes [History of Qumes]*. Tehran: Kāviān.

Hamedan Meteorological Organization (2017). Hamedan's climate. URL: www.sinamet.ir.

Hamidi, J., & Hamidi, M. (1997). *Bar Farāz-e Ghollehā-ye Alvand Az Ālmābelāgh Tā Kolāh-ghākhi [On the top of Alvand from Almabelagh to Kolah-ghakhi]*. Hamedan: Maftoon-e Hamedāni.

Hassouri, A. (2014). Knocking on Bowls. In F. Hashabeiky (Ed.), *International Shahname conference: the second millennium* (pp. 87–91). Sweden, Uppsala: Uppsala University.

Hatefi Shoja', S., & Zareyi, M. E. (2017). Mostanad-sāzi Va Arzyābi-ye Bāzsāzi-ye Bāgh-hāye Tārikhi-ye Hamedān Bar Asās-e Manshoor-e Florence [Documentation and evaluation of the restoration of Hamedan's historical gardens according to florence convention]. Bāgh-e Nazar, 14(46), 31–40.

Hedayat, R. G. K. (1871). *Fahang-e Anjoman-Ārā-ye Nāseri [Anjoman-Ara Persian dictionary]*. Tehran: Ketāb-forooshi-ye Eslāmiyeh.

Hojjat, I. (2006). Tasir-e Eslāhāt-e Arzi Bar Shekl-e Roostāhā-ye Irān [Impact of the land reform on the Iranian villages' morphology]. *Honar-hāye Zibā, 26*(1983), 75–84.

Ildermi, A. (2012). Barresi-ye Elal-e Tashkil Va Voghooa'-e Jaryān-e Vārizeyi Dar Dāmane-hāye Shomāli-ye Alvand Hamedān [Causes of the formation and occurrence of debris flows at the Northern Base of Alvand]. *Fazā-ye Joghrāfiyāyi, 12*(37), 217–245.

Javadi, S. (2013). Bāz-khāni-ye Revāyat-e Āb Va Derakht Dar Dowrān-e Eslāmi [A review on the ancient story of water and tree in the Islamic period]. *Honar Va Tamadon-e Shargh, 1*(1), 43–50.

Khalaf-e Tabrizi (1938). Farhang-e Borhān-e Ghātea' [Borhan-e Ghatea' Persian dictionary]. Sherkat-e Taba'-e Ketāb.

Khatibi, A. (2015). Barresi-ye Tasir-e Har Yek Az Moalefe-hāye Sāzgāri Bar Ensejām-e Khānevade-ye Aghvām-e Shahr-e Hamedān Bā Rooykard-e Jāmea'e-shenāsi [The impact of adaptation factors on the cohesion of Hamedan's ethnicities from a sociological point of view]. *Majale-ye Jāmea'e-shenāsi-ye Irān, 16*(2), 135–169.

Labbaf Khaneiki, M. (2006). *Water division systems in Iran (Persian)*. Tehran: Iran National Water Museum.

Labbaf Khaneiki, M. (2019). *Territorial water cooperation in the central plateau of Iran*. Switzerland: Springer.

Mohammad Padeshah, (1957). *Farhang-e Ānendrāj [Anendraj Persian dictionary]*. Tehran: Khayyām.

Mokhles, F., Farzin, A. A., & Javadi, S. (2013). Mazār-e Pir Morād; Manzar-e Farhangi Āyini-ye Shahrestān-e Bāneh [Pir Morad Tomb; cultural and religious landscape of Baneh Township]. *Nazar, 10*(24), 27–38.

Molla-zadeh, K., & Taheri Dehkordi, M. (2011). Tārikhche, Jāyābi Va Sākhtār-e Hegmatāneh-ye Mādi [History and structure of the Median Ecbatana]. *Fasl-nāmeh-ye Motālea'āt-e Shahr-e Irāni Eslāmi, 6*(2011), 5–16.

Momeni, M., Kajbaf, A. A., & Allahyari, F. (2013). Māhiyat-e Turkamanān-e Arāghi Va Rābete-ye Ānhā Bā Turkamanān-e Saljooghi [Historical root of Iraqi Turkmens and their connection with Seljuq Turkmens]. *Pajoohesh-hāye Tārikhi, 49*(2), 19–36.

Motarjem, A. (2011). Neshāne-hāyi Az Yek Mohājerat-e Ghomi Be Shomāl-e Gharb Va Gharb-e Irān Dar Hezāre-ye Sevvom Dar Parto-e Kavosh-hāye Bāstān-shenāsi [Indications of an ethnic migration to the Southwest and West of Iran in the 3rd Millennium BC in the light of archaeological studies]. *Nāmehā-ye Bāstān-shenāsi, 1*(1), 137–146.

Moyin, M. (1972). *Farhang-e Fārsi-ye Moyin [Moyin Persian dictionary]*. Tehran: Amir-Kabir.

Nafisi, A. A. (1964). *Farhang-e Nafisi [Nafisi Persian dictionary]*. Tehran: Khayyām.

Nasir Khusraw, (1978). *Divān-i Ash'ār-i Hakim Nasir Khusraw Qubādiyāni [Nasir Khusraw's poetry book]*. Tehran: McGill University in cooperation with University of Tehran.

Noorbakhsh, M. R. (1987). Sāa't-e Mekāniki Dar Irān [Mechanical clock in Iran]. *Āyandeh, 13* (6&7), 395–408.

Rudaki (2017). Remnants of Rudaki's Poetry; Kalila and Dimna, Sandbad-Nameh. Visited on 28/04/2017, http://ganjoor.net/roodaki/masnaviha/kalila-sand/sh6.

Shafayi Hamedani, S. H. Divān-e Kāshef al-Hayvah.

Wah, T. K., & Shukri, Z. (2013). Memes and consciousness: friend or foe. *Social and Behavioral Sciences, 97*(2013), 517–521. https://doi.org/10.1016/j.sbspro.2013.10.267.

Yekta, S., Mohammadi, K., Farahani, S., Yekta, L. (2014). Naghsh-e Qanavāt Dar Shekl-giri-ye Sākhtār-e Fazāyi-ye Shahr-hāye Irāni-Eslāmi [Role of qanats in the formation of the spatial structure of Iranian-Islamic cities]. Second International Conference on Construction, Architecture and Urban Development.

Zareyi, M. E. (2011). Sākhtār-e Kālbadi-ye Fazāyi-ye Shahr-e Hamedān Az Āghāz-e Dowre-ye Eslāmi Tā Pāyān-e Dowre-ye Qājār Bar Asās-e Madārek Va Shavāhed-e Mojood [Spatial structure of Hamedan from the beginning of the Islamic period to the late qajar period based on the existing documents and evidences]. *Nāmeh-hāye Bāstān-shenāsi, 1*(1), 57–82.

Chapter 3
Hydro-political Organization

Abstract In the Iranian lands, sizable cities with a considerable number of inhabitants could not be organized haphazardly. Such a population established an organic nexus of socio-economic relationships among them—necessary for their integrity—through a political body. This political body was able to sustain the population by attending to water affairs that were vital to the pervasive agricultural economy, in the shape of supplying water and protecting against floods. This political body in fact invested in water affairs whose higher efficiency resulted in better agriculture and accordingly more tax. Therefore, water systems have always been a source of income for Iranian governments whose hydraulic agenda systematically tethered the peasants to the cultivated lands. This chapter examines the interrelations between water resources and political organization from a historical–geographical point of view, which have taken different shapes in river-based and qanat-based economies over the course of history. At least since the Sassanian Empire, governments have had a systematic dependence on a river-based economy nourished by irrigated cultivation. Hence, governments used both bureaucratic and military tools to secure their control over water as the vital sources of revenue. However, governments' interests were not limited to water only as source of finance, but they were also concerned about water to strengthen their political ties with their preferred territories through water allocation systems. Within their kingdoms, some hydro-social territories were given primacy in terms of water shares, through both intensive hydraulic constructions and water management systems. Hydro-social borders were delineated by the political organization through water allocation based on political, religious and social priorities, and those borders altered from time to time following political changes. Such governmental interventions in water affairs could lead to inequitable water allocation, based on a socio-political pattern that seems to have prevailed over the Iranian history. This chapter also shows how "water management" could set the stage for the collapse of those political organizations, when it outgrew the governments' internal capacities. A systematic relationship between water and political organization suggests a historical model that helps explain the rise and fall of many Iranian dynasties.

M. Labbaf Khaneiki, *Cultural Dynamics of Water in Iranian Civilization*,
https://doi.org/10.1007/978-3-030-58900-4_3

3.1 Introduction

King Enki was God of water (Jacobsen 1987, p. 455) who built the temple of Abzu to bless and patronize the Sumerian city of Eridu, near present-day Basra (Kramer 1972). This mythology is a manifestation of our argument that the God-like kings could unite and stabilize the society in Iran and Mesopotamia by means of water.

This chapter mostly focuses on the Iranian plateau, but it is impossible to talk about the Iranian history without pointing out Mesopotamia that used to house the headquarters of at least two Persian empires for centuries.

Although Mesopotamia enjoys two permanent rivers, the supply of water could not keep up with the growing demand of cultivated lands (Wilkinson 2003, p. 76). From 3000 BC onward, water resources have constantly been on the decline (Nissen and Heine 2009, p. 46), but cultivated lands tended to expand. Dwindling water supply brought about two main consequences: First, sporadic patches of land were increasingly left without direct access to water resources, leading to greater consternation of the population in the principle cities (Ibid, p. 47). Second, water shortage gave rise to more centralization of power that crystallized mostly around water resources management. Although during Uruk period (3500–3100 BC) a locally administered water sector—based on chiefdoms—began to evolve into a more centralized water management (Wright 1977), large-scale water supply systems came into existence in the Kessite period (1531–1155 BC) for the first time (Wilkinson 2003, p. 92). Such extensive water supply systems could not be constructed and maintained without the sponsorship of state authorities (Ibid; Malayeri 1996, p. 111).

Similarly, in many parts of the Iranian plateau, irrigation paved the way for a massive agriculture that took advantage of the fertile sediments that seasonal floods had settled over thousands of years. Artificial irrigation led to a boom in agricultural production, which made it possible for human populations to grow to the extent that was unprecedented by them (Kuiper 2011, p. 18). Such a large population—concentrated in a certain area—and the sprawling irrigation networks were harmonized through political organizations that intervened in water management, levied tax on agricultural products and safeguarded hydro-social borders. These functions of political organizations could in turn sustain Iranian societies for many centuries. This chapter examines a "cycle of prosperity" in which Iranian states sponsored development of irrigation, then development of irrigation attracted and sustained larger population, this larger population provided more workforce and more workforce resulted in even further development of irrigation.

Government's intervention in water allocation systems served the interests of political organization in two manners. First, water could provide the political organization with the financial sources required to perpetuate its main functions, through agricultural economy based on irrigation (Wilkinson and Rayne 2010). Second, water contributed to the sovereignty of political organization through devising "hydro-political borders".

A historical interplay between the social origin of the ruling power, ethno-ideological hegemony and geopolitical concerns separated different regions along sort of hydro-political borders. Sometimes, these borders triggered hydro-social tensions across Iran and Mesopotamia (Swyngedouw 2004, p. 150; Budds et al. 2014, p. 167), first of which broke out between Lagash and Umma city-states over their rights to Tigris River some 4500 years ago (Dinar 2007, p. 22). The ancient governments have always manipulated natural distribution of water resources to the detriment of some communities but to the benefit of some others, based on the political agenda and power relations. For example, in the Sassanian period, a portion of the Tigris was conveyed by irrigation channels which lay on the Persian or eastern side of the river. But after Sassanians' fall, another system of canals was derived from the western bank of the Tigris, which transferred water to the north of western Baghdad (Le Strange 1900, p. 48).

Nevertheless, Iranian civilizations have not been homogeneous in terms of water–power relations. The Iranian plateau is bounded in the north by Alborz Mountain Range and in the west by Zagros Mountains, where most of the permanent rivers originate and then pour into the Caspian Sea, Persian Gulf or some inland lakes and swamps. Hence, a vast area in the central plateau is bereft of any permanent surface stream, which historically set the stage for the formation of a different economy based on groundwater extraction through the indigenous technology of qanat. We can say that river-based economy came into being in Mesopotamia and around the Iranian plateau that enjoyed permanent rivers, whereas qanat-based economy was associated with an arid geography that prevailed over the central plateau.

3.2 Water and Political Power

Agriculture has been the main source of the governments' revenues—the same role as that of industries nowadays—and has always been reliant on irrigation and water allocation systems that resulted from technological achievements, institutional organization and political decisions. Hence, water management used to come within the purview of the Iranian sovereignties, at least in the river basins where powerful trans-regional governments came into being. It is undeniable that water was not alien to the political systems in other civilizations. For example, the Aqua Virgo (built in the first century BC) proves how an aqueduct served as a vehicle for driving political agendas in Rome (Hunter 2020, p. 26). But sheer dependence on water affairs was a trait that distinguished Iranian political systems. Water could potentially provide the state with the budget required to maintain its political and military capabilities, provided that water was processed and economically enriched by agricultural activities (Wilkinson and Rayne 2010).

In a variety of references, it is very prevalent to see something about the role of large-scale bureaucratic irrigation systems in the history of Iran and Mesopotamia (Somervill 2010, p. 24; Rzoska 1980, p. 23; Ur 2014, p. 9; Wittfogel 1981), of which Wittfogel's theory is well known. According to Wittfogel, the technical progress of ancient hydraulic societies led to a kind of development "trap" in which the social and political power embodied in the bureaucratic control of water was used to defeat social change (Stroshane 1988, p. 124). Therefore, the nature of water management in Iran could bring about oriental despotism in two ways. First through the organization of a large population which could provide the intricate irrigation system with the required labor, and second through the accumulation of power which was rooted in the crucial role that water played in the production systems under such geographical conditions (Wittfogel 1981). However, some parts of this theory have been shattered by many controversies over the past decades (Lees 1989; Peet 1985; Mitchell 1973; Millon et al. 1962; Hunt et al. 1976, p. 390).

Although nowadays Wittfogel's theory may be seen as anachronistic determinism in the present scientific circles, some contemporary scholars are still reminiscent of its principles. Wittfogel's idea still inspires those who are interested in how power is materialized through a relationship between physical and political structures (Strang 2016, p. 292). For example, Worster (1985) says "control of water meant wealth and power, while lack of control meant relegation to the lower classes, struggle for existence, and a lack of individual choice". According to Worster, construction of hundreds of dams and water transfer pipelines in the American West during the twentieth century is an industrial resurrection of ancient water-controlling societies, which has turned the arid West into the hydraulic West (Mauch and Zeller 2008, p. 5). Even many empirical researches done by Wittfogel's critics suggest that a relationship between irrigation and political rule was likely to exist, but did not necessarily lead to an authoritarian political system in form (Bichsel 2016, p. 360). Radkau takes it for granted that hydraulic infrastructures and political power have always been intertwined over the course of history, though he excludes their relationship as a simple causality. He thinks about interaction between water and power in terms of possibilities. An ambitious government vigorously tends to expand the irrigation systems as far as possible as a way of increasing the tax revenue (Radkau 2008, p. 91).

There are a great deal of historical evidences that show how much the Iranian political fabric has correlated with water management affairs, a correlation that contributed to the integrity and sustainability of urban populations. Iranian plateau was ruled by the Sassanian dynasty from 224 to 651 AD (Elman 2004, p. 95), when the central government apparently intervened in the rapid expansion of a costly irrigation system (Frye 2006, p. 132). Hydraulic systems, whether for irrigation or for providing water to medium and large urban sites, can be seen as a key component of the centrally planned economic developments in Sassanian socio-economic sphere (Wilkinson et al. 2012, p. 171). Hole (2007, p. 518) highlights "irrigation system" as a starting point for understanding the Sassanian history.

Water issues were so important for Sassanian kings that the administrators who lacked enough knowledge about water were shunned as incompetent (Malayeri 2000, p. 57), and this same tradition continued even into the time of Ghazzali (1058–1111 AD) who believed that an understanding of groundwater and its extraction is the knowledge highly required in the royal secretaries (Rajabzadeh

2005, p. 135). Sassanian king Khusrau I has been quoted as saying "royal power rests upon the army, the army depends on finance, finance on taxes, taxes on [irrigated] cultivation, and cultivation on justice" (Yarshater 2006, p. 398; Jankowski 2016, p. 20). Land cultivation has been contingent on irrigation and water across those arid and semi-arid territories, and king Khusrau's word subtly highlights an existential relationship between the kingdom and water.

Although the water division system of Zayandeh River in Isfahan is believed to be invented by Sheikh Baha'i (1547–1621 AD) (Mahmoodian and Qayoomi Bidhendi 2012, p. 144), it actually dates back to ancient times, probably when the first agricultural settlements began dotting along the river bank and then competing over the available water. Ibn Rustah (1986, p. 183) has mentioned Ardashir (180–242 AD), the founder of the Sassanian Empire, as the first person who regulated the water division of Zayandeh River. The Sassanian kings usually gave primacy to hydraulic structures, of which Qaysar Dam is notable, built by the Sassanian king Shahpur I (241–272 AD) near Shushtar in Khuzestan (Khazraee 2008). Another historical example of the governmental intervention in building hydraulic systems is a water-dividing dam in the town of Damqan, reported by Abu Dulaf the Arab traveler who attributed the dam's construction to the Sassanian kings. "Damqan enjoys a strange dam from the Sassanian period, built to distribute water. Water gushes out of a mountain cave, where it is divided among one hundred twenty villages for the purpose of irrigation. One hundred twenty canals derive from the dam to transfer those water shares to the beneficiaries, and in all canals the amount of water never changes, and it is not possible to merge them either. It has a wonderful structure and I have never seen any dam better than or even similar to this dam in other regions" (Abu Dulaf 1975, p. 81). Water and its management were considered so crucial that the tax was calculated based on the amount of water available each year (Khosravi 1969, p. 49).

In the early thirteenth century, Hamavī Baghdādī (2004, p. 41) defined tax as "the peasants' income that should be delivered to their lord". Caliph Umar (584–644 AD), as quoted by Hamavī, has said that tax is the price of safety for those who choose to live in the lands of the Islamic caliphate (Ibid). In many cases, tax was imposed on the defeated and occupied territories not in return for civil services by the triumphant government as we perceive in our modern societies, but only in return for the right to live. If a particular territory was conquered after a battle, its subjects had to pay much more tax than those who had surrendered without bloodshed. Hakim al-Nishāburi (933–1014 AD) tells a story in his book, which bears witness to the fact that the amount of tax had something to do with the way the subjects submitted to a ruler's authority, of course coupled with other factors. He says that Abdallah ibn Tahir (798–844 AD), the Tahirid governor of Khurasan, intended to calculate all water sources and all cultivated areas in order to establish a new taxation, after he succeeded in bringing the whole of Nishapur region under control. Ahmad ibn Haj Bakr, one of the Nishapur's dignitaries, advised Abdallah ibn Tahir by saying that "keep in mind that you did not conquer our city in battle, but our city fell into your hands peacefully. Hence, you can get your share of our harvest just the same amount as the former rulers did. This portion of our harvest as traditional tax has long been in place in

Nishapur, which we always paid to the rulers, and its amount has been recorded by all tax collectors. If you are content with this same amount of tax, we will follow too. Otherwise, get out of our city and give us some time so we can reap our crops and store up foods for wartime and barricade ourselves in our city fortification. Then come back and fight us. If you can win the battle, you have free rein to levy tax however much you wish" (Hakim al-Nishāburi 1996, p. 207).

Anyway, this tax should have usually been proportional to the total harvest expected to be yielded at the end of the farming season, and the amount of irrigation water proved to be a clear indicator to predict the amount of coming harvest. The amount of water-based tax varied from time to time and place to place with political situation, hydrological condition and agricultural capacity. Sassanian king, Kavadh (473–531 AD) charged the peasants one-forth, one-fifth, one-tenth or sometimes one-twentieth of their crops for annual tax, based on their soil fertility and water availability. The closer a farmland was to a water source, and the more arable its soil, the bigger portion of the harvest belonged to Kavadh (Bal'ami 1999, p. 679; al-Athir 1992, p. 107).

Sassanian Empire set up an irrigation bureau as well, which managed all affairs related to water division and ownership to ensure the tax that water shareholders were supposed to pay. The same bureau was inherited by the Arab conquerors in the seventh century under the name of Divan al-Ma', which made this region one of the richest in the known world, according to Le Strange (1905, p. 3). The people responsible for water affairs were selected from among adept officials who were called Qayyas or Hassab in the water administration system at the time of Abbasid caliphate (Papoli Yazdi and Labbaf Khaneiki 2001, p. 49). The hydraulic legacy of Sassanian Empire continued into the Muslim dynasties that later ruled over Mesopotamia and the Iranian plateau.

The book "al-Kharaj" gives credence to the fact that governmental taxation was largely dependent on water affairs in the Abbasid period. Abu Yusef (731–798 AD), its author, contends that an efficient tax system entails a capable water supply system under Islamic ruler's supervision (Abu Yusef 2004). Another book with the same title, authored later at the time of Caliph al-Muktafi in the early tenth century, makes the same statement about water (Katib Baghdadi 1991). Many rulers had to heed such recommendation, like Adud al-Dawla (ruling from 949 to 983 AD) who repaired and dredged the entire network of canals in Baghdad (Souli 1935, p. 477). As an instance, there are many classical history books showing that the political organization was always under apprehension of both drought (Ibn al-Jawzi 1995, p. 70; Ibn al-Athir, 1994b, p. 529) and flood (Ibn Kathir 1998, p. 586; Ibn al-Athir 1994a, p. 443; Ibn al-Jawzi 1995, p. 77; Zahabi 1998, p. 24; Ibn al-Fuwati 2002, p. 146), which could potentially undermine their economic structure.

Those in charge of water affairs were called Moqassem al-Ma' at the time of Ghaznavids (977–1186 AD) in the city of Merv, and Qavvam or Hafazeh in the city of Nishapur (Papoli Yazdi and Labbaf Khaneiki 2001, p. 49). Those officials were actually part of tax system, who tried their best to keep the agricultural activities in order, and regulate such important production factors as water mostly to the benefit of the ruling power (Bartold 1971).

Water-based taxation was not only practiced in Mesopotamia and Iran, but it was also common in Egypt whose economy was anchored in irrigated horticulture. Ibn Battuta (1304–1369 AD) mentions a Nilometer or a structure devised for measuring the Nile River's annual fluctuations regarding taxation. Every year in June, the water level was measured in the hope that it came up 16 zera' or about 16.5 m, which boded well for the ruler who could receive his agricultural tax in full according to plan. The water level between 16 and 18 zera' was still a bonanza that brought more harvest and accordingly more tax, but after 18 zera' they expected a devastating flood to flush away their crops and shatter their economy. A water level less than 16 zera' was considered a bad omen for the coming year when the shrinking farmlands did not afford the ruler's tax (Ibn Battuta 1997, p. 75).

This correlation between water level and tax in Egypt has been mentioned by al-Maqdisi too earlier in the tenth century AD. In 847 AD, the Abbasid caliph wrote to his governor in Egypt to rebuke him for the small amount of tax that he had collected from the farmers. The caliph expected the Egypt tax to be much more than the amount delivered by his governor, but it turned out to be less than even that of his poorest provinces at the time. al-Maqdisi interprets this story as a misunderstanding that was caused by the unpredictable nature of the Nile River. According to him, it was impossible to expect those Egyptians to pay a constant amount of tax, since their agriculture was fully dependent on the changing hydrological regime of the Nile. Hence, their tax had to be calculated as a flexible percentage of their final harvest, which was proportional to the amount of water available in the Nile. However, al-Maqdisi contends that the farmers of Levant and Khurasan were not exempt from the constant tax as he writes that "unlike Egyptians, those farmers could not evade the government's tax with the excuse that water shortage did not allow them to reap the same amount as before, but whoever reaped less was just lazier" (al-Maqdisi 1982a, p. 94).

Sometimes the governments invested in water supply systems and irrigation structures to help expand the cultivated lands that yielded more tax. In the Samanid period (819–999 AD), the region of Bukhara was traversed by 12 water canals all derived from Zerafshan River. All those canals except the canal of Khetfar were built by people on the kings' orders. Narshaki (899–959 AD) highlights those irrigation canals as precious sources of income over which many battles broke out between rival powers. Every canal had a certain tax system according to which the farmers were annually charged by the ruler (Narshakhi 1984, p. 45).

Taxable waters were called kharāji, in which the ruler's right was inherent. Whatever land irrigated by kharāji water became subjected to the same tax (Hamavi Baghdadi 2004, p. 46). In other words, a taxable flow of water transferred its financial property to wherever it ran, provided it was used for irrigation. Fuqahā or Islamic jurists used their expertise to bestow legitimacy on the measures taken by the ruling powers in terms of water-based taxation. Their fatwa (ruling) portrayed any objection to the water tax in a way that it was tantamount to heresy. For example, Ibn Abī Laylā (638–702 AD), Abubakr bin Abī Sabra (657–719 AD), Abdullāh bin Zakwān (685–748 AD), Abū Hanīfa (699–767 AD), Malik bin Anas (711–795 AD), Sofyān bin Saeed (716–778 AD), Zufar ibn al-Hudhail (728–774 AD), and Beshr ibn

Ghayyās al-Marisi (755–834 AD) were all influential clergymen who gave their verdicts on the water tax by asserting that "if a man leaves his land uncultivated, he must be ordered to irrigate and cultivate it and then pay its tax, otherwise his land must be confiscated and handed over to someone else to do so" (Al-Balādhuri 1958, p. 625). The holy men's apparent enthusiasm for the irrigation issues was not specific only to those Islamic jurists, but it was present in every political system that hinged on collusion between the men of sword and the men of temple rather than on public acceptance. They lived in a vital symbiosis that ensured the integrity of religious foundations on the one hand and the legitimacy of political systems on the other hand. That was why many of the Zoroastrian scriptures left from the Sassanian period were awash with recommendations particularly about irrigation like dam construction, qanat digging, canals maintenance, and so on (Christensen 1989, p. 431), and that same tradition continued into the Islamic period.

The measures taken by Tahir, the founder of Tahirid dynasty (821–873 AD), well exemplify the significant role of ruling powers in the water management systems, when he managed to bring together the clergymen from all over Khurasan to pool their legal knowledge and compile a book on water division laws (Salimi Moayed 2000, p. 158). A similar intervention has been reported from the time of king Malek Moez al-Din Hossein in the fourteenth century. He was from Kart dynasty who ruled over a vast area of Khurasan for about 148 years, and their kings were based in the city of Herat in present-day Afghanistan as their capital, where Heray River flowed. After a dispute broke out between water shareholders in Herat, Malek Moez al-Din drew up a set of laws regarding water division. That same law served as a legal basis on which the river water was divided among the cultivated lands in Herat for a long time (Amiri 2001). Noelle-Karimi (2014, p. 15) explains how units of administration corresponded to the irrigation system in Herat. Such interventions in water management systems ended up benefiting the governments most. If there was no regulation in water consumption, most of the farmers' energy was bound to be put into the conflicts and competitions over more share of water, and as a result, the agricultural activities became bogged down to the detriment of the governments.

However, sometimes in Iranian history the irrigated agriculture was not robust enough to satisfy the government's existential requirements, for example due to a drought or the increasing rapacity of the king's tax collectors or both. In this case, the government resorted to "plunder economy" by invading other territories, the same role that the governmental extraction of exhaustible resources—like oil—is playing nowadays. For example, Mahmud (971–1030 AD), the Ghaznavid king, became more and more reliant on plunder economy, as he invaded India and looted its wealth at least 16 times (Salehi and Amir Shekari 2013, p. 163). Political and ideological factors—like building up more religious legitimacy at the Abbasid Sunni Court—served as a starting point in the Ghaznavid excursions to the Indian subcontinent, where Mahmud could exhibit a resemblance between his ardent campaigns and the Islamic primitive holy wars against infidel pagans. Ghaznavids' battles in India drained a huge budget to the extent that even the lucrative agricultural lands in Khurasan no longer afforded to pay their costs, though the kings generously paid their soldiers in cash (No'pasand and Isavi 2013, p. 106) and gave free rein to their commanders and many of their soldiers to take a big chunk of the booty in every battle. Beyhaqi (1971, p. 142) reports that "in one of the wars waged by the Ghaznavids, the king bestowed on his soldiers most of the booty including gold, silver and slaves. The king claimed only the best weapons and then divided the rest of the booty among his army". Thus, the Ghaznavids spilled into a vicious cycle that led to their collapse ending their relatively short reign. The frequent costly battles with India and other territories could raise the Ghaznavids' earnings in the short term, but it undermined their political integrity in the long term. The Ghaznavids had to put more pressure on the farmers of Khurasan to pay more tax, most of which was allocated to their army and holy battles. The farmers gradually became disillusioned with the insatiable demand of the tax collectors, and lost their motivation to stay involved in agricultural activities, most of whose revenue was usurped from them as tax (Bosworth 2007, p. 171). Although the Ghaznavids could return from their battles laden with booty, most of that wealth had to be spent on the same army and the next military adventures rather than on such sustainable income sources as agriculture. Some of the farmers viewed those battles as a business more lucrative than toiling away on the farmlands, and joined the Ghaznavid garrisons, contributing to the formation of plunder economy, and at the same time an increasing downturn in agricultural sector.

Although water management and division system were the cornerstone of agricultural economy across the Iranian plateau, they did not usually remain unchangeable in a particular place. Population growth, a change in the region's cropping pattern, new immigrations into the region and a rise in the number of water users could all bring about considerable changes in the water management system. For example, migration of Arab tribes to the town of Qom and their increasing involvement in agricultural activities gave rise to a longstanding conflict between them and the local farmers, which eventually led to a change in the region's water management system. This conflict has been reported in the book "History of Qom" written in 988 AD, which bears witness to the outstanding role of ruling powers in water management and water allocation (Qomi 1982, p. 47).

It was not only political organization that could affect water affairs, but sometimes water affairs could also exert influence on political organization. For example, Merv, once one of the principal cities of historical Khurasan, enjoyed an intricate water management system. The farmlands of Merv used to be irrigated by Murghab River, which originated from central-western Afghanistan. The water share of each farmland was precisely calculated and recorded by a bureaucratic immense organization linked to the central ruling power. al-Maqdisi (946–991 AD) reports that the most powerful commander along with 10,000 soldiers presided over Murghab dam to ensure its proper function, and his political power surpassed the king's (Meftah 1992).

It seems that Iranian lands have not been homogeneous in water–power relation. Different types of water supply systems seem to be important variables to explain how water and political organization have been correlated. Therefore, such grand theories like Wittfogel's, which try to generalize about the entire Iranian land, do not seem very tenable, when we look for an explanation about the disparity between river-based and qanat-based economies.

Wittfogel has omitted a vast area in the central plateau of Iran, whose political economy has been anchored in groundwater having been extracted by qanat systems. In a qanat-based economy, no large-scale bureaucratic water management system came into existence, a system that is said to pave the way for the emergence of trans-regional monarchies. Hence, Wittfogel's theory probably holds true only when it comes to river-based economies, where the huge volume of water and the vast lucrative cultivated lands were associated with an extended water management system with a powerful economic and political nature (Labbaf Khaneiki 2019, p. 40). Failure of Wittfogel's theory to generalize about the entire Iranian lands has something to do with different scales of analysis (Hunt et al. 1976, p. 390). While talking about river-based economies, the debate is on the scale of trans-regional sovereignties. But when it comes to qanat-based economies, the scale changes to rather local. A geographical–historical analysis on local scale does not necessarily end up in the same result as that of national or trans-regional scale.

In qanat-based economies, water-state relations did not bring about the same political dynamics as stipulated in Wittfogel's theory. Decentralized water management and endogenous water cooperation were inherent in qanat-based economies (Labbaf Khaneiki 2019). However, in river-based economies, water-state relations crystallized around governments' concentration on water affairs. Therefore, one of the pivotal elements of the river-based economies is governmental intervention in water affairs. There are plenty of historical records that give credence to governmental intervention in water affairs mostly in river-based economies. Those records show how much Iran's political fabric and their water management affairs have been interwoven, which is manifested in the Iranian political structure even today.

However, the structural peculiarities of qanat gave rise to a different economy that diverged from river-based economies in terms of "water-state relation". The water management system of qanat was bereft of any top-down intervention. In a qanat-based economy, a spontaneous water management was in place to regulate water supply and demand. The qanats were managed mostly by petty landowners based in sporadic agricultural oases where political power could not be concentrated. Those in charge of water division were selected by the qanat shareholders, and could be easily ousted by a unanimous vote when needed (Labbaf Khaneiki 2019). But, why did not qanat-based economies accommodate the same political structure as that of river-based economies?

To answer this question, first of all we should look into the physical peculiarities of qanat system that did not demand a huge deal of human labor, despite the considerable length of many qanats, which sometimes reaches 80 km. A qanat whose length reaches tens of kilometers has not been built over night, but it is the result of centuries of people's struggle to keep groundwater running in their qanat gallery. Every qanat was only a few hundred meters long in its infancy, but its tunnel was gradually advanced into the aquifer only a few meters a year in response to the groundwater drawdown in the wake of climatic fluctuations. The inhabitants kept extending the qanat tunnel to maintain at least the same water flow, as long as the groundwater drawdown did not overtake their financial and social capacity. Otherwise, they abandoned the qanat and moved to somewhere else favorable for building another qanat. A typical qanat was always in balance to a particular population, their available resources and labor. That population grew to the threshold of the qanat capacity to sustain their own economic lives, if they could successfully retain the same water discharge. Hence, the qanat water was sufficient to sustain a subsistence economy at best, which was far from appealing to such political organizations. The qanat capacity to bring water was limited to its physical features, and its water could not be increased in a way that a vast area could come under cultivation to produce a good surplus for the tax collectors. Iranian ancient governments relied on the economic resources of their headquarters most, though they counted on the tax collected from their other provinces as well. If great governments chose a qanat-dependent town as their seat of power, it was very challenging for them to properly organize their fiscal structure based on a subsistence economy that could hardly sustain even the local communities. Hence, they had to become over-reliant on their subjects' revenue that came from other provinces, which was potentially an existential threat to their sovereignty. A simple rebellion or political turmoil in their rich provinces could totally obstruct their financial channel and then topple the whole system. If we look at the history of the qanat-dependent regions in the central plateau of Iran like Yazd, we do not find any trans-regional dynasties successful in subjugating a territory as vast as that of for example Sassanians, Abbasids, Ghaznavids or Seljuks.

The qanat supplied limited water, and the population who lived around it was too dynamic to become a reliable ground for developing an extended political system. Unpredictable discharge of qanat caused the inhabitants to develop industrial skills that were deployed to supplement their income deficiency when the qanat water dwindled. Qanat system paved the way for a diverse economy in which the local industries played a significant role. The local industries were contingent on human skill rather than on land, and human skill gave a more mobility to the local communities whose toil was supposed to fill the king's coffer. The farmers were tethered to their lands as their source of income, whereas the craftsmen made a living mostly by their skills that yielded the same result wherever they went. Later in this chapter, I will explain that the concept of territory pertained to the demographic quality rather than the geographical definition of a particular area of land, from the perspective of Iran's historical governments. Hence, an area whose population—or in fact source of income—was always prone to leave did not lend itself to becoming the seat of a lavish government.

3.3 Hydraulic Collapse

I mean by "hydraulic collapse" a situation when socio-economic and political structures fall into an irreversible disorder that is in the first place sparked by water resources crisis. Although water issues stand out as the initial cause of such a collapse, it actually unfolds within a mesh of different factors that amplify each other's impact to a large extent. From a historical standpoint, in a society whose revenue was deeply anchored in irrigated agriculture, a disproportionate taxation could have a devastating impact on the entire socio-economic system. Ibn Khaldun (1332–1406 AD) was one of the first scholars who noticed a subtle relationship between inflation of power, growing taxation and political collapse. I will return to his notion about hydraulic collapse later on the following pages. Ibn Khaldun's notion is partly echoed by Marvin Harris in his book where he builds upon Wittfogel's work to clarify the interplay between irrigation and despotic political systems. Harris takes a step further and comes up with a historical model that tries to explain the life cycle of all oriental dynasties, based on the pivotal role of hydraulic systems. Most of the oriental dynasties follow the same historical destiny that begins with their earnest investment in irrigation infrastructures and water management. The further a dynasty develops hydraulic systems, the more lands would come under cultivation, and the more tax would pour into its coffer. The dynasty's income gradually grows out of proportion with its administrative capability to regulate such fiscal affairs and properly re-invest its income surplus. Hence, "water money" sets the stage for a cancerous corruption that leads the entire political system to the abyss of a multifaceted bankruptcy. As a result, "public works are neglected, the dikes begins to leak, the canals fills up with silt, and production declines" (Harris 1977, p. 175). The dynasty becomes so weak that it eventually gives way to the opportunistic invaders or even homegrown rebellions. The next dynasty sets in, and repeats that same vicious cycle.

Historical evidences testify the fact that disproportionate water taxation resulted in the depletion of social capital and natural resources both. The farmers did not afford to quench the government's thirst for more tax that was constantly demanded to keep pace with its growing bureaucratic and military ambitions. More pressure on the farmers took the shape of more pressure on soil and water resources to the point that the farmers lost their economic motivation. For example, centuries of extensive irrigation in Mesopotamia drastically changed the landscape, and brought about soil salinity, sedimentation and soil degradation that gradually led to more decline in cultivation efficiency in the course of time. The process of soil degradation that culminated in such severe issues as soil erosion, nutrient loss, and shortage of organic carbon was triggered by agricultural intensification (Wilkinson 1997, p. 76). The Diyala River was utilized for irrigation at its full capacity even before the end of the Parthian period, let alone under the Sassanian dynasty whose fame rested on its ardent enthusiasm in developing agriculture. During the later Sassanian and the early Islamic periods, the entire cultivable area was brought under cultivation, and the expansion of irrigation system was about to exceed the

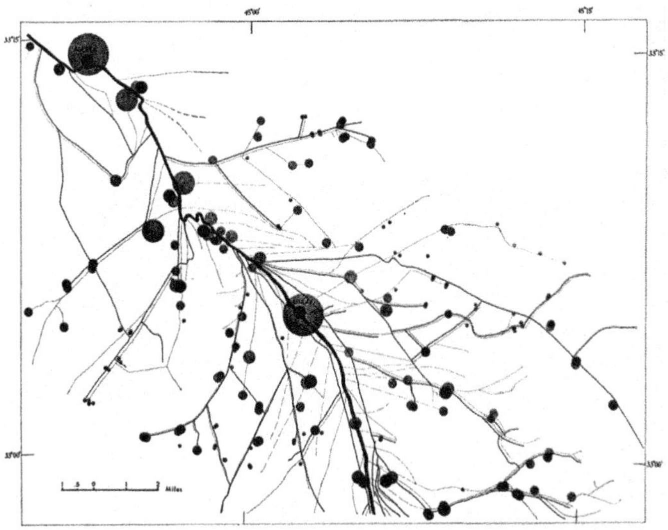

Fig. 3.1 Extensive transverse canal system along the Nahrawan great canal in Mesopotamia, which was built and developed during the later Sassanian and the early Islamic periods (Jacobsen and Adams 1958, p. 1256)

natural limits of the region's water resources and sometimes did so (Jacobsen and Adams 1958, p. 1256). In Mesopotamia, intensive irrigation ended up in a high water table that exacerbated soil degradation and salinization in the lowermost alluvial plain, which is believed by some scholars to have been one of the driving forces behind the fall of the Sassanians (Butzer 2012, p. 3635) (Fig. 3.1).

Although many of the farmers still had no choice but to scrape a living by squeezing the exhausted water and soil resources, corruption was another factor that germinated under the shadow of an unhealthy interaction between the government and its subjects. Hence, the farmers became more disenchanted, since most of their wage was confiscated either as tax or as bribe by the corrupt tax collectors. Under the Ghaznavid Sultans (977–1186 AD) it became rife that the tax collectors charged the farmers something for their own personal profit, and the farmers always had to pay more than determined as official tax (Yusefi 2008, p. 259). Ghaznavid Sultans also practiced a special tax farming system based on fief called Iqta‘ according to which a particular area was temporarily granted to a commander for a period of time, where he had a free rein to extort as much tax as he wished directly from the subjects, in return for his military service (Erfanian 2014, p. 11). Moreover, the Sultan was very likely to issue promissory notes called barāt for his soldiers as well, addressed to the same Iqta‘-holder or his tax officer there, which put an additional strain on the subjects. Barāt was a substitute for a soldier's official salary, which sometimes proved more profitable than what he was to get in cash from Sultan (Ibid, p. 261). Many of barāts were cashable only up to a certain amount of money mentioned in the document, by ‘Āmel who was responsible for the taxation of that particular area, especially during the period of Saljuqs in the late twelfth century AD (Yusefifar and Azadeh 2013, p. 56). In every fiscal area, all tax collectors worked under ‘Āmel’s supervision, who was usually appointed by the Iqta‘-holder. Such financial system paved the way for a systematic corruption that in the end took its toll on the public and the government alike.

Among the Iran’s dynasties, Safavid dynasty is a clear example, whose increasing dependence on water systems harbored corruption that contributed to their fall in the early eighteenth century AD. Dependence on agricultural tax became the cornerstone of the Safavid political economy from its very beginning (Aghili 2016, p. 177). The king’s concern for water management is quite understandable, because in a hydraulic civilization like what we see in the Safavid Isfahan, the kingdom’s stability, authority and sovereignty closely correlated with an effective water management system. Majority of the then population were involved in agricultural activities that proved impossible without irrigation. During the dry months when the crops were in need of more water, the amount of water was disproportional to the water demand of the extensive cultivated lands whose revenue not only sustained the local economy but also propped up the ruling power economically in the form of taxation. Every year the royal officials estimated the amount of water expected to be available until the end of the cropping season, as an indication to predict the amount of harvest on which the annual tax could be levied. Chardin (1643–1713) the French traveler reports that the farmers of Isfahan had to pay a specific water tax in addition to the tax they were regularly charged for their crops at harvest time. They were obliged to pay 20 sols[1] for each jerib[2] of land as a tax levied on the river water, but the tax of spring water was cheaper (Ravandi 2003, p. 156). Thus, any disorder in the water management could wreak havoc on their agricultural production systems and accordingly could bring down the entire economic structures and then the government itself. An impaired water division and management system could bring about a chaos in the society where many farmers became ensnared in a ring of violence rather than focusing on their agricultural production. As a result, the widespread social dissatisfaction and a downturn in the government’s revenue could pave the way for its fall.

The water division of Zayandeh River was supervised by the king himself. For example, in 1611 AD when a drought struck the region, Shah Abbas I (1571–1629) put a ban on rice cultivation in Lenjān area to save the water share of those who lived in the river downstream. Although the King was personally into the water management of Zayandeh River, this responsibility was usually entrusted to a Mirāb, an official in charge of water issues under the king’s direct supervision. All over the country, Mirāb was appointed and ousted by the king at his discretion. The story that happened to the Mirāb of Mashhad at the time of Shah Abbas II (1632–1666 AD) well testifies to the high status of water management at the Safavid court.

[1]20 sols were equal to 1 livre in French currency. 6 livres were equal to 1 dollar in both Halifax currency and Quebec currency (Mccullogh 1983, p. 86).

[2]In Iran Jerib was a unit of land measurement, which was equal to 1000 m^2.

A number of Mashhad's dignitaries sent a petition to Shah Abbas II against Ali Beyg, the city Mirāb, who allegedly misused his position. The king immediately summoned Ali Beyg along with all petition signatories to investigate their complaint and make a judgment. The problem was so important for the king that he decided to play the role of judge himself, and eventually ruled that Ali Beyg be ousted and replaced by another official named Beyg Arestāv (Qanzvini 2004, p. 641).

The position of Mirāb was ranked among 21 top dignitaries who had special permission to attend the royal assembly in the king's presence, and Mirāb was the twelfth person. Tavernier the seventeenth-century French traveler reports that "in Safavid Isfahan, Mirāb is responsible for all waters. This official position is among the highest governmental posts in terms of its benefit. One can compare this post to the directorate of waters and forests in France" (Tavernier 1990, p. 578). The post of Mirāb was very lucrative and coveted by most of the officials. A Mirāb's salary was up to 4000 tomans a year (Ravandi 2003, p. 157), whereas for example the king's special physician was given an annual salary of about 340 tomans at the time, and other royal physicians were paid much less; only 28 tomans a year (Danesh-pajooh 1968, p. 300). Mirāb's income was not limited to his regular salary that he received from the king every year, but he was entitled to a major portion of the gifts that his officials overtly got from the farmers (Ravandi 2003, p. 157). Such a treasure was handed to Mirāb in return for his services as follows: 1—to appoint the chief of each water canal; 2—to supervise the annual cleaning of all canals; 3—to ensure the proper division of Zayandeh River among all districts; 4—to prevent any conflict over water between the shareholders; 5—to settle any dispute over water between the farmers (Ibid; Afshar 2001, p. 575). Hence, such a position as Mirāb who had full control over the most important production factor in an agrarian society had a predisposition for corruption, to the detriment of the entire political system.

The water management system of Zayandeh River, which was run by Mirāb appointed by the Safavid kings, was increasingly afflicted by corruption and nepotism at the time of Sultan Suleiman I, and gradually drove the Safavid dynasty into annihilation at the time of Sultan Husayn, along with other historical causes. Engelbert Kaempfer (1651–1716), the German explorer who visited Isfahan at the time of Sultan Suleiman I, reports that "given that the crops are fully reliant on the water being rationed among the farms, it is very common for the water officials to take bribes, and the desperate farmers vie with each other to appease the officials" (Kaempfer 1981: 104). As a result, the economy spilled into a dark epoch, eventually leading to the collapse of Safavid dynasty. Chardin describes Iran at the time of Sultan Suleiman I as follows: "this year (1676) Sultan Suleiman is still reigning. Just over a short period, only twelve years since the previous king, a downturn in economy has occurred so drastically that the country's wealth seems to have declined by half" (Chardin 1966a: 39). Sultan Suleiman I turned a blind eye to the oppression of the farmers at the hands of his Mirāb. Chardin believed that Sultan Suleiman I was more lenient with the offending Mirābs than were his predecessors, as he writes in his account of travel. "In 1676, the farmers of a vast area located

42 km from Isfahan complained to the king that his Mirāb had charged them three thousands livres[3] in return for water to irrigate their farmlands for ten days. However, Mirāb broke his promise and gave them water only for one day. The farmers brought with them a leafless branch of a dead tree as a testament to the tremendous damage that Mirāb's betrayal inflicted on their orchards and farms. The king heard their heartbreaking plight and then only slightly penalized Mirāb, whereas such an offence could lead to a capital punishment if it happened at the time of any other king. This king usually goes easy on the officials who violate people's rights and the king's reaction would not be more than fining the offender or changing his post at most. Nevertheless, the money extorted from poor people is rarely returned to them, since its three forth has been already delivered to the king's treasury. However, when it comes to the king's personal rights, the offender would be summoned while chained and shackled, and he may be put to death" (Chardin 1966b: 1180).

In the later Safavid period, Mirāb's coffer brimmed with the money the desperate subjects had to pay just to make ends meet. Sultan Husayn (1668–1726 AD) gave his Mirāb a free rein to charge the subjects a considerable amount of money for their own water shares, apart from what they officially paid as tax (Afshar 2001, p. 575). Thus, many farmers fled Isfahan in droves to evade the unbearable demand of the government for agricultural tax, despite a law that came into force at the time of Sultan Husayn banning any migration from agricultural lands without government's permission (Salim 2018, p. 112). Nevertheless, only in Isfahan region more than 1000 villages were completely abandoned and their farmlands were left uncultivated, shortly before Mahmud Hotak conquered Isfahan and finished off the Safavid dynasty in 1722 AD (Ibid, p. 113).

Almost no dynasty was spared from water corruption that eventually contributed to their political disintegration, since the tax levied on irrigated agriculture was the backbone of their economy. The last historical examples of water corruption pertain to Qajar dynasty that ruled over Iran from 1789 to 1925 AD. Henry-René D'Allemagne who visited Iran in 1911 reported that the region of Bakhtiari ran short of water in the summer when the cultivations were at the peak of their need for irrigation. Mirāb took advantage of such a high demand for water and accepted bribes to give more water than written in the water-right documents, in the face of all efforts that the village chief made to oversee the water division (D'Allemagne 1956, p. 1044). Around the same year, Kamareyi writes in his diary as follows: "I am fed up with this chaotic government. No water has reached our gardens over the last fifteen days. I always pay Mirāb's fee upfront, I stoop to flattering him and I tip him extra money, but in vain. Every week I look forward to the water that Mirāb has promised to deliver, and every day I stay at home lest Mirāb comes over to fulfill his promise, but he never shows up. I fear that things get worse if I complain to the government or municipality about Mirāb's shortcoming. Mirāb is treating me this way as unfairly as possible, let alone I provoke his antagonism by complaining

[3]Chardin has converted the then local currency into French livre.

about him. Mirāb does not do his job properly, even if I luck out and eventually get my water. Three or four parts of my garden are always left dry, since Mirāb hastily cuts off the water before I complete my irrigation. If Mirāb is not assured of the government's support and if they do not act in collusion, he dare not commit such crimes" (Kamareyi 2005, p. 291).

As mentioned, Ibn Khaldun elaborates on a historical model that explains the rise and fall of many oriental dynasties. He says that "disheartened peasants can undermine a strong government only through their farmlands lying idle, which leads to a fatal decline in the government's revenue" (Ibn Khaldun 1996, p. 542). According to Ibn Khaldun, most of the dynasties arise from a Bedouin background, where they are accustomed to a Spartan lifestyle. When they come to power, they begin developing their bureaucratic organization that enables them to earn enough budget needed to strengthen and maintain their costly army (Ibid, p. 539). A strong army in turn ensures a ceaseless current of tax mostly imposed on the agricultural areas, and such accumulation of wealth sets the stage for an extravagant aristocracy. At this stage, the government tends to expand its sumptuous court by hiring more dispensable staff, building more unnecessary departments, and recruiting more lavish dancers, musicians and concubines. Now the government has to satisfy the insatiable appetite of its growing aristocracy, by pressuring the peasants to cultivate more lands and give more tax without caring about their rights. The government has to increase its pressure on the peasants in order to keep pace with the growing demand of the royal court, to the point that the peasants eventually flee their homes and abandon their farmlands that are the cornerstone of the government's economy. At this stage, the government suffers a dramatic bankruptcy that is associated with a downturn in agricultural production. The government's asymmetrical growth turns into the cause of its own demise, when it reaches the point that it no longer affords to fulfill the expectation of its sizable aristocracy (Ibid, p. 538). In the absence of enough budget, the government enters a state of political cannibalism that is marked with multiple competitions and conflicts between different parties along the faults of power within the royal circle. Hence, the government cracks and breaks down into several independent rival dynasties, or it may succumb to foreign invaders lurking behind its borders. Ibn Khaldun does not recognize the commoners toiling away on the farms as an immediate factor in the rise and fall of such dynasties like what happens in a widespread peasant revolt, but he takes into consideration their indirect impact on the government's disintegration by hampering the tax system (Fig. 3.2).

The historical records are replete with facts and examples that substantiate Ibn Khaldun's theory. The rulers vied to get the most tax, as if they dealt with insensate objects attached to the farmlands, whose only mission in life was to fatten the government at their own cost. Caliph Umar (584–644 AD) levied a tax of 4 dirham on each jerib of wheat, 2 dirham on each jerib of barley and 8 dirham on each jerib of date palm, which brought him a total revenue of 128,000,000 dirham a year only from Mesopotamia. However, al-Hajjaj (661–714 AD), as governor of Mesopotamia, had no qualms about imposing the highest tax on the farmers. al-Hajjaj could raise the government's income to a total amount of 18,000,000,000 dirham only through a taxation reform in Mesopotamia (al-Maqdisi 1982a, p. 158).

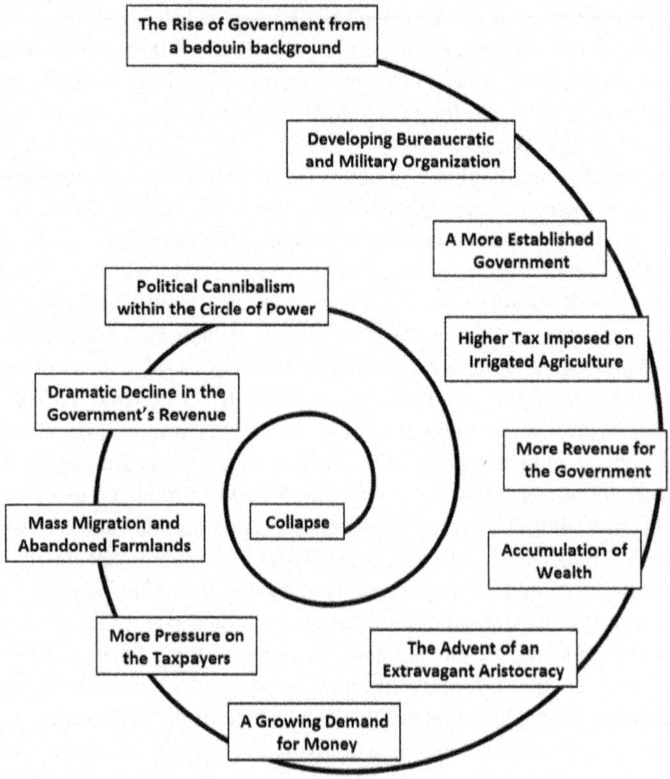

Fig. 3.2 Ibn Khaldun's theory on hydraulic collapse in a nutshell

This money dwarfed the tax obtained from the Iranian plateau at the time, as Gorgan annually paid 10,196,800 dirham, Qumis paid 1,196,000 dirham and Beyarjomand paid only 26,000 dirham to the Umayyad Caliphate (al-Maqdisi 1982b, p. 549). Although the Umayyad Caliphate was flooded with such a huge amount of money, their income always lagged behind their growing expenditure, to the point that the increasing tax imposed on their subjects led to the devastation of many agricultural infrastructures like irrigation systems. The Abbasid Empire eventually took the place of the Umayyad Caliphate, and tried to repair the agricultural infrastructures that were in ruins after years of oppression. The recovery of the irrigation system was in full swing under Harun al-Rashid (786–806 AD), but began to deteriorate again as a result of "wasteful expenditures and rapacious tax farming" (Butzer 2012, p. 3635). Thus, the Abbasid Empire went down the same path of that of the past governments.

With the government's growth, its expenses soared owing to its military development that was deemed inevitable to ward off external invasions and internal rebellions. Moreover, the growing aristocracy foisted an additional cost on the government, which entailed more pressure on the agricultural taxpayers who were considered the most important financial source. For example, at the time of Mahmud (971–1030 AD), the Ghaznavid Sultan, the farmers of Khurasan bore the brunt of any rebellion that broke out in the country, since they had to go along with a fresh taxation that Mahmud repeatedly imposed on them to finance his campaigns (Bosworth 2007, p. 164). Before one of those battles, Mahmud ordered to raise the money he needed to mobilize his men and weaponry only in two days. His order was immediately carried out by fleecing the farmers (Bartold 1987, p. 607). Those costly battles were not the worst, but the lavish aristocracy drained out a great portion of the governmental revenues. Building gorgeous monuments was of great interest to the oriental rulers who wished to exhibit the majesty of their kingdom, as a historical and geographical milestone. Sometimes the rulers sacrificed their other interests only to build a monument like the mosque of Damascus, on whose construction the tax of Syria was totally spent for 7 years by Al-Walid ibn Abd al-Malik, the sixth Umayyad caliph (674–715 AD). "The mosque of Damascus is one of the world wonders. Its floor is thoroughly paved with white marble, and the beautiful paintings on its huge walls are so vivid that one may mistake them for real animals and plants" (Qazvini 1994, p. 246).

In the tenth century, the Abbasid caliph swapped the total yearly tax of Bukhara for some opulent clothes made of an extremely expensive fabric named Shādorvān, as Narshakhi (899–959 AD) reports. "Bukhara enjoyed a textile workshop located between Hisār and Shahrestān, in the vicinity of its congregational mosque. The workshop was unique in weaving a fine Shādorvān and then in making very sumptuous clothing that was afforded only by the caliph. Every year, the caliph's tax collector was dispatched to Bukhara to take the aforementioned clothes as the region's tax" (Narshakhi 1984, p. 28). Hence, 100% of the Bukhara's tax was spent on the caliph's personal finery, and none of it was used to improve the subjects' living condition. Even though a ruler gave of a share of his revenue back to his subjects, his generosity was usually targeted at renovating or developing the water supply systems that could later pay off in the shape of more production and then more tax.

Basra was initially built as a garrison town in 636 AD by the Muslim conquerors in order to settle Arab soldiers and their dependents (Hoyland 2016, p. 49). In its formative years, Basra's population exceeded 250,000, including 80,000 soldiers (Kennedy 2011, p. 177), who quickly turned to agricultural activities to sustain such a growing population. Basra turned into an economic hub, which exported its five main products, including dates, grain, wool, hides and horses to different regions (Simpson 2016, p. 36). In the face of Basra's wealth, its people themselves struggled with poverty. Ali ibn Abi Talib (601–661 AD) says in Nahj al-Balagha (the collection of his sermons, letters and utterances) that "Basra has the filthiest land among all God's cities, and it is the closest city to water but furthest from heaven" (Kavandi and Shad-del 2013, p. 101).

Basra was supplied with a brackish water that was of poor quality for both drinking and irrigating, when Abdullah ibn Umar (died 750 AD) paid a visit to the city. He decided to have a new canal built for Basra, but one of his officials objected that the project would cost as much as 300,000 dirham. Abdullah replied that "if the canal would consume even the total yearly tax of Iraq, I still would not back down" (Al-Balādhuri 1958, p. 517). The canal was eventually built, named after Abdullah ibn Umar, but it benefited the government rather than the subjects in the end. About two centuries after the canal was built, al-Maqdisi (946–991 AD) reported that in Basra, fresh drinking water was still at a premium and potable water was imported from the region of Ableh by ship, which was then affordable only by the upper social class. "The water of Basra is not drinkable, since one third of it comes from sea and the rest is contaminated by sewage, and only the poor desperate people drink it" (al-Maqdisi 1982a, p. 180). Although the dense population of Basra was of great benefit to the government by providing more workforce to the agricultural sector, it begot more pollution, more diseases, less access to safe drinking water and deteriorating living conditions for the commoners. Therefore, the subjects were doomed to live in dire poverty though they were the lucrative source of the government's wealth. Al-Maqdisi admires Mesopotamia for its fertile soil, prosperous agriculture and legendary wealth despite the poverty that blighted the mass of people in the tenth century AD. "Mesopotamia is the land of prosperity and wealth, which is a great favorite with caliphs, scientists, poets and clergymen. Basra is on a par with the entire world, and Baghdad is coveted by everyone on the earth. Its river is ceaseless as if it is coming from heaven. However, those cities are teeming with miserable people all in the grip of expensive prices, conflicts, and squalor. Everyday their situation worsens and they endlessly suffer from unbearable taxes, poverty and prostitution" (al-Maqdisi 1982a, p. 157).

When the government's rapacity crossed the tolerance line of the subjects, different backlashes began to challenge the ruling power. Migration was a common backlash, which could cripple the production system and its ensuing taxation. What al-Heravi (died 721 AD) says about the city of Herat under the third king of the Pishdadian dynasty, Tahmuras, well portrays the Iranians' mentality about the relationship between their rulers and people, which in the end results in the people's exodus and the rulers' fall. "The city of Herat was founded by Tahmuras, son of Hooshang. Tahmuras lasted long in power, and his might and majesty grew beyond imagination, until his arrogance drove him to the Satan's trap. He claimed to be God and bedeviled his subjects by demanding more and more tax that was far beyond the tolerance of the beleaguered peasants. Tahmuras filled up his own coffer with the money of those desperate people to the point that they eventually fled their lands in droves and spread all over the world" (Al-Heravi 2004, p. 64). However, migration was not the only consequence of the overtaxing of farmers. Sometimes, they showed their discontent by defecting to the king's enemies, when they no longer afforded the rising tax. In the early eleventh century AD, while Mahmud was busy splurging his money on the Indian battles in Multan, Qarakhanid tribes seized the opportunity and occupied Nishapur that was the hub of irrigated agriculture in the Ghaznavid territory at the time. Nishapur's fertile soil, abundant water and elaborate irrigation system (Ibn Rustah 1986, p. 200; Estakhri 1994, p. 270; Ibn Hawqal 1987, pp. 166–168; Mustawfi 1983, pp. 226–228; Fouache et al. 2011) gave rise to a lucrative agriculture such that only a portion of the agricultural income of a village in Nishapur was equal to the total revenue of the city of Merv at

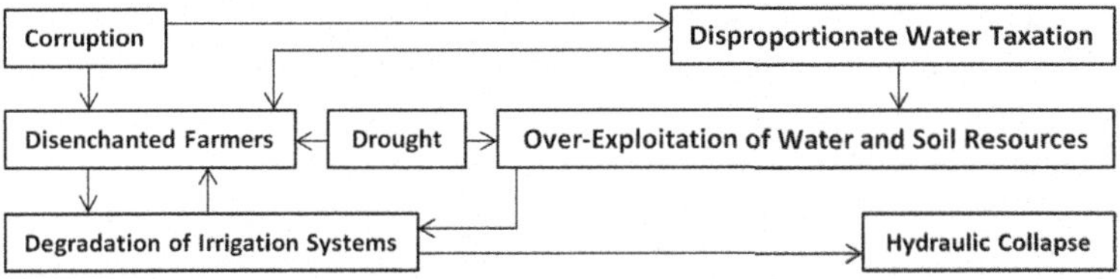

Fig. 3.3 Ecological and socio-economic factors whose interaction leads to hydraulic collapse

the time (al-Maqdisi 1982a, p. 438). But majority of Nishapur's farmers who had become disillusioned with the rapacious tax-collectors defected and easily opened the city's gates to the invaders (Bosworth 2007, p. 171) in the hope that the invaders would usher in an era of relief. The process leading to hydraulic collapse has been summarized in the following model, which befell almost all Iranian dynasties to some extent before the modern oil economy changed the traditional equation in the early twentieth century (Fig. 3.3).

3.4 Hydro-political Territories

Hydro-political territories may be subsumed under the category of hydro-social territories. Hence, I prefer to begin this topic with the comprehensive concept of "hydro-social territory" that helps us better understand the government's intervention in water allocation systems. Hydro-social territory is referred to as an area of land where there exists an intricate interplay between social dynamics and water resources. Hydro-social territories were not necessarily congruent with hydrological boundaries like watersheds or qanat-catchment areas. Boelens et al. (2016, p. 1) define hydro-social territories as socially, naturally and politically constituted spaces that are (re)created through the interactions among human practices, water flows, hydraulic technologies, biophysical elements, socio-economic structures and cultural–political institutions.

From a historical point of view, on the Iranian plateau and even in Mesopotamia, water resources were usually scarce in proportion to water demand. Hence, it was very crucial for any government to allocate water resources based on its political priorities. The way water resources were allocated to different territories could sustain or undermine a political organization even at national level. In Iran, some hydro-social territories were deprived of their traditional water shares to the benefit of others, through governmental intervention. In fact, some territories expanded their hydro-social boundaries into that of others with government's backing, though the hydro-social borders altered over time in the wake of political changes. The important question is that who was given the water at whose cost? To answer this question, we should look into the power relations that played a crucial role in prioritizing the country's regions, when it came to allocating the available water resources. There was a complicated interplay between the social origin of the ruling power, ethno-ideological hegemony and geopolitical concerns, which drew invisible borders between the regions, having given rise to hydro-social tensions across Iran. In general, such hydro-social cycles were infused with a myriad of social tensions and were contested terrain at each moment of its flow (Swyngedouw 2004, p. 150), and were also expressive of politicized nature of water management over the course of history (Budds et al. 2014, p. 167).

Under Sassanian Empire, some irrigation canals were derived from the eastern side of the Tigris to transfer water to the Persian communities who were brought from the Iranian plateau in order to unbalance the region's ethnic composition in the king's favor. But after Sassanians' fall, the eastern canal system fell by the wayside and was replaced by another system branching off from the western bank of the Tigris, conveying water to the north of western Baghdad (Le Strange 1900, p. 48). Nevertheless, the Sassanian kings did not allow both residential and agricultural units to clump together in the vicinity of their capital city Ctesiphon, probably as a defensive strategy. Instead, they tended to establish extensive irrigational infrastructures in the Diyala area as their financial backing, far away from Ctesiphon that was in the frontline of any potential clash. The landscape of Diyala clearly reflects the intensive irrigation practiced over the past millennia (Verhoeven 1998, p. 167). Perhaps, they had learned from the fate of Seleucia with a population of about 80,000, which was reduced to rubble by Avidius Cassius in 165 AD, and its irrigation system did not survive the ordeal (Adams 1965, p. 70). The Sassanian kings considered a strategic dichotomy between their political center in Ctesiphon and economic hub in Diyala, in order to reduce their vulnerability to Roman attacks. Otherwise, those irrigation structures remained an easy target for the invaders, whose destruction could bring down the entire economic and political systems. This way they expanded the hydro-political border of Ctesiphon as a survival strategy to better withstand the possible foreign invasions.

Under Sassanians, Persian colonization of Mesopotamia was underlain by the development of water supply systems. Shapur I (died 270 AD) founded the town of Hasar-Shapur that accommodated ethnic Persians entrusted with considerable political and military control. The same colonization policy was pursued by Shapur II (309–379 AD) who brought 12,000 Persians to Nisbis (Christensen 2016, p. 105), and water supply was an inevitable prerequisite for such policies. Shapur I built the Nahr Shayla Canal for the colonists whom he settled in the village of Misikhe, north of Fallujah. The Nahr Shayla Canal provided enough water to sustain much larger population of the colonists to the extent that Misikhe turned into a large city renamed Anbār (Ibid, p. 103). Anbār or Hanbār is a middle Persian word in origin (Faravashi 1992, p. 41), contrary to Christensen who knows it as Arabic. Anbār means "storehouse" or "depot" in Persian, which implies its logistical significance for Shapur I during his attacks into Syria.

Shapur I also founded another city named Gondeshapur in the present-day province of Khuzestan, southwest of Iran (Simpson 2017, p. 28). Archeological surveys reveal that the city layout follows a network of tunnels delivering water from a canal off-take from the River Dez (Ibid, p. 30). Building new canals and transferring water to different locations seems to have been a ceaseless project during the Sassanian time. One of the most important canals was called Nahrawan whose construction is ascribed to Khusrau I (512–579 AD) (Dashti 2005, p. 43). Nahrawan branched off from the Tigris and meandered southward, irrigating an area extending over 100 km to the present town of Baquba (Mays 2008, p. 472) where it discharged its water into the Diyala River (Dashti 2005, p. 43). The Nahrawan Canal along with its levees, dams and side branches turned Mesopotamia into a legendary treasure for the Sassanian kings at the time. That was why they called this part of their country "Dil-e Iranshahr", which meant the heart of Iran. The Nahrawan Canal retained its economic and political status even after the Sassanians until the mid-tenth century AD, when it began to silt up and deteriorate, which heralded the impending collapse of the Abbasid caliphate (Adamo and Al-Ansari 2020, pp. 188 and 189).

The Sassanians were a clear historical example of a government that took advantage of the fluidity of water to enhance their political dynamism. In Mesopotamia, most of the early Ubaidian and Sumerian villages and towns were built very close to the rivers (Nardo 2007, p. 314), where they lived and cultivated both. Hence, their hydro-political territory did not go far beyond their settlement and its nearby cultivated area. However, water is liquid and could be transferred easily by gravity to further areas whose natural and geopolitical conditions were best suited for agriculture. Thus, later many levees were built to divert the river water to an extensive network of artificial canals, which made it possible for the Sassanians to bring a vast area under cultivation as the remote economic hinterland of Ctesiphon. Many Persian farmers loyal to the Sassanian king were settled across that hinterland to which the governmental canals network supplied water, and a network of trading routes took their products back to the capital city (Akihiko 2003, p. 176). The hydro-political territory of Ctesiphon grew larger and stretched around all cultivation areas irrigated by an extensive network of canals devised by the government. The agricultural production units were spread out across a vast hydro-political territory in a relatively safe zone, behind the urban facade of Ctesiphon. Water was allocated to loyal shareholders who were expected to ensure a ceaseless flow of wealth into the royal treasury. The water allocation system in fact mirrored the power relations, which successfully sustained the Sassanian dynasty for more than 400 years.

Therefore, a hydro-political territory was bounded by a dynamic border that was determined mostly by human populations and their irrigation-based production systems. Both population and production system were dynamic, and as a result a border delineated based on them became dynamic too. The Iranian empires were bereft of any contiguous territory, as we see in our modern border system. For example, the Sassanian territory is called "network empire" (Lawrence and Wilkinson 2017, p. 119) that consisted of sparse patches of imperial control interspersed with spots lacking notable political organization (Diener and Hagen 2012, p. 35). Those ancient empires presided over relatively vague borders that were quite different from our modern-day notion about border and territory. Our modern border system traces its origin to the European colonialism (Ibid, p. 37) that valued the lands not only for their available human labor but also for their raw materials like minerals and later oil sources strewn across the natural landscapes. However, the ancient political paradigm viewed "territory" as the location of tax-payers and the extent of the king's jurisdiction rather than the physical geography of the country's boundaries. In other words, borders pertained to people rather than geographical places. In general, it was people who could work on such raw materials as water and soil to produce an income surplus that underlay the government's economy, whereas mountains, deserts, forests, lakes and seas had no value on their own except when a road traversed them. Even the roads were also dependent on the people living along them, who had to bear the burden of providing food and fodder for the soldiers and their horses while being deployed to a battle on the frontier (Pourshariati 2008, p. 316). That supply was not usually included in the subjects' annual agricultural tax. Hence, even the most brutal rulers did not want their subjects to disappear from their domain, unless the ruler was more of a passing invader with no certain political agenda who had not come to establish a state but to sack the settlements and return.

The founder of the Achaemenid Empire, Cyrus (600–530 BC) advised his commanders that they should not treat their subjects with contempt, since they are the source of the kingdom's wealth. "You should treat them such that they always stay in their settlements, because a densely populated country is the rich one, and a thinly populated country is devoid of wealth" (Briant 2005, p. 116). Achaemenid kings placed a high value on human population whose toil on the farmlands was regarded as the key to their government's wealth, success and prosperity. They encouraged population growth by giving rewards to the families who had more childbirth every year (Ibid, p. 114). A larger population was on a par with a greater, richer and stronger country, in their opinion. For that reason, some of the ancient kings were very concerned about the quantity of their population rather than its quality. For example, at the time of Peroz I (died 484 AD), many Iranian towns and villages were struck by a severe drought that lasted for 7 years. Many rivers, springs and qanats ran dry and a persistent famine blighted a large part of the Sassanian territory. Apart from Peroz's prayer at the shrine of "Adar Khara" for an end to the drought (Pourshariati 2008, p. 363), he ordered the affluent to share their food with the poor. He even threatened that "I will punish the entire population, if I ever learn that any single person has starved to death in any village or town across the country" (Ibn al-Athir 1992, p. 298). By our today's standards, it was hardly fair that a government shifts the responsibility of supporting hungry people to other members of the public, whereas the onus must have been on the government to disburse the aid at least under such urgent circumstances. But in the past, it was considered a fair measure that a ruler probably took in order to save the number of his workforce in the time of famine. Some rulers were also likely to reduce taxes or even temporarily exempt some groups of their subjects from paying the tax, when the subjects were on the verge of escape or fatal starvation. The rulers who were not too hard on their subjects in the time of famine were titled and recorded in history as "dispenser of justice", an epithet given to king Khusrau I (512–579 AD) for example (Dinawari 1992, p. 100). Those measures were not taken out of generosity but as a political strategy to retain members of such a hydraulic society as the crucial elements of the kingdom's territory.

A hydro-political territory was more of a fiscal nature than geographical. It pertained to the communities who paid homage to the central government mostly by rendering a part of their agricultural revenues to the royal tax collectors. When a community refused to pay tax, it was tantamount to losing a part of the government's territory. It was the common mentality of all Iranian historical governments that "territory" was not really bounded by such natural features as useless mountains, ravines or seas, but was more determined by the locations of people who had the capability to transform water into precious products through agriculture and then pay their tax to the central government. For this reason, in Iran, geography as a field of science was always considered an important topic about places but inseparable from fiscal facts. The Iranian geographers had to describe the governmental-fiscal affiliation of each town and village in order to figure out the territories, and it was not possible for them to do so without knowing to whom those places paid their yearly tax (Ibn Khordadbeh 1889, p. 18). For example, when Ibn Fondoq (1097–1169 AD) describes the geography of his homeland, he repeatedly uses fiscal terms to specify different locations in relation to each other and to the main territory. "Jājarm belongs to the region of Joweyn, whereas Khār is part of Beyhagh territory, because its tax is included in the yearly tax of Beyhagh" (Ibn Fondoq 1982, p. 34).

The city of Kabul was annexed to the Abbasid territory since its tax was delivered to al-Ma'mun (786–833 AD), and later it simply separated from the Abbasid territory as soon as its tax was diverted to the Kharijites (Markwart 1994, p. 194). A territory was subdivided into smaller territories according to the government's fiscal organization. For example, under Nasr ibn Ahmad (906–943 AD), Qumis and Nishapur were two main fiscal territories that both constituted a great part of the Samanid hydro-political territory. Nevertheless, different territories sometimes overlapped, and their positions did not follow any regular geographical pattern. For example, at the time, the region of Beyarjomand was considered part of Nishapur territory, because its tax was delivered to Nishapur's tax bureau. But there was a village in the middle of Beyarjomand, whose tax went to Qumis, and its inhabitants regarded themselves as people of Qumis unlike their surrounding villagers who felt belonging to Nishapur (al-Maqdisi 1982b, p. 549).

Therefore, we should not project our own perception of "border and territory" onto the past. Before the Iranians adopted the European concept of territory with all its aftermath like modern nationalism, territory was more of a demographic-fiscal nature. Such historical territory is understandable in a discourse whose rules can be analyzed outside the time in which it has occurred (Foucault 2002, p. 220). The government's dependence was principally placed on the agricultural revenues of its own immediate location, and the remote locations were of secondary importance. The locations whose inhabitants paid tax were recognized as parts of the central government's territory whose identity hinged on the people irrigating and cultivating. The peasants who lived in the remote parts of a particular territory had a different notion of central government from what we do today. In their vision, government was probably embodied in the first ring of the taxation chain, which was a local tax collector who showed up every year at harvest time to collect the agricultural tax. Territory was perceived as a fiscal hierarchy that started at the level of villages and ended at the royal treasury (Labbaf Khaniki 2012). The governments wished to rule over the people who could create wealth through irrigation rather than coveted unpopulated barren lands. That was why those historical territories were more of a hydro-political phenomenon that proved to be a crucial component of the political organization.

3.5 Conclusion

In a river-based economy, considerable amount of water within a particular area of land favored mass cultivation over the course of Iranian history. A political body supported by such an economy took it for granted that more lands under cultivation would result in more tax. Regardless of how those political bodies came into existence on the basis of irrigation in the first place, almost all of them tended to expand irrigated agriculture that brought them more wealth. Development of irrigation was contingent on two main factors: "irrigation management organization" including all mirābs, water officials, tax collectors, canal workers, dam builders, accountants, etc., and "agricultural workforce" that was supplied and renewed through constant migration, colonization and childbirth. The more the irrigated agriculture developed, the more people flocked to both management organization and working units. Hence, a centralized political body was needed to organize such a large number of employees involved in the irrigation management on the one hand, and to prevent or settle conflicts in the agricultural units on the other hand, which could otherwise bog down the entire system. Although the political body's tendency to become more despotic led to a higher income from the agricultural units, the influx of money eventually turned into the leading cause of political disintegration. Such higher income set the stage for an asymmetrical development where the bureaucratic and military entities eventually outgrew the production infrastructures. At this stage, higher income gave rise to more items of expenditure, to the point that even the maximum tax on the agricultural units no longer sufficed to cover the government's cost. Irrigation tax gave birth to military adventures, splendid palaces, costly aristocracy and extended bureaucracy, which uncontrollably grew into a monster that swallowed its mother in the end. Such growing expenditures drained the government's revenue part of which should have been spent on the irrigation infrastructures in order to keep the economic system working. But first of all, the subjects at the base of the power pyramid bore the brunt of the deteriorating and shrinking irrigation infrastructures, because the government usually continued to impose the same tax in order to maintain its own costly extended structure. Any attempt to prune the government in order to reduce costs could lead to a political turmoil. Hence, most governments found it easier to exploit the taxpayers, though the exhausted agricultural units no longer afforded to meet the government's growing demand. A disenchanted workforce could continuously lower the government's income, which culminated in the disintegration of its hydro-political territory. Given that the main element of a historical territory was its inhabitants whose labor was translatable into wealth, any considerable demographic changes like outmigration, fatality or defection could be interpreted as its demise.

In a qanat-based economy, the relationship between water, agricultural units and government took place in a different way. Qanats were dispersed water resources whose small flows could not be merged into a great stream in order to irrigate a vast integrated area of farmlands, due to topographical obstacles. At least once a year a particular qanat required cleaning and repair to keep running. The small dimensions of a qanat tunnel did not have room for more than between 4 and 6 workers for dredging or digging underground, who were provided by the same beneficiary community. A particular qanat was usually shared amongst between 100 and 1000 shareholders who regulated their relationship based on a spontaneous face-to-face acquaintance. The relatively small number of both qanat workers and shareholders and the personal relationship between them did not lend itself to the formation of a centralized polity. A normal qanat discharged only between 15 and 150 L of water per second, which varied from year to year with the amount of precipitation. Hence, the qanat water was not reliable enough to attract a great deal of workforce that could in turn become a necessity of devising a centralized management. In contrast to the river-based economy, in the qanat-based economy, water management was of an endogenous nature, which was handled by all shareholders, and then resulted in a more cohesive community. Their agricultural economy had to be supplemented by local small industries that actually imported water from other areas in the shape of virtual water. Such industries stepped up social mobility that did not favor centralized polities.

References

Abu Dulaf. (1975). *Abu Dulaf's travel book (Persian translation).* (A. Tabatabayi, Trans.). Tehran: Zavvar.

Abu Yusef (2004). *Kitab al-Kharaj [Book of Tax].* (M. A. A. Shoaa', Trans.). Kabul: Meyvand Publisher.

Adamo, N., & Al-Ansari, N. (2020). The Abbasids and Tigris irrigation canals: the Nahrawan. *Journal of Earth Sciences and Geotechnical Engineering, 10*(3), 183–222.

Adams, R. M. (1965). *Land behind Baghdad; a history of settlements on the Diyala Plains*. USA: The University of Chicago Press.

Afshar, I. (2001). *Majmoo'e Asnād Va Manābe' Tārikhi [Collection of historical documents and sources]* (Vol. 1). Tehran: Bonyād-e Moghoofāt-e Doktor Mahmood-e Afshār.

Aghili, S. A. (2016). Nesbat-e Hākemiyat-e Siyāsi Va Hayāt-e Eghtesādi Dar Dowre-ye Avval-e Hokoomat-e Safaviyeh [The relationship between political ruling and economic life in the early Safavid period]. *Tarikh- Eslam Va Iran, 25*(28), 171–213.

Akihiko, Y. (2003). Urban-rural relations in early eighteenth century Iran. In K. Nobuaki (Ed.), *Persian documents; social history of Iran and Turan in the fifteenth to nineteenth centuries* (pp. 147–185). London: Routledge.

al-Athir, A. (1992). *Al-Kāmil fī al-Tārikh; Tārikh-e Bozorg-e Eslām Va Irān [The complete history; great history of Islam and Iran]*. (A. Hālat & A. Khalili, Trans.). Tehran: Moassese-ye Matbooā'ti Elmi.

Al-Balādhuri, A. I. Y. (1958). *Futūh al-Buldān [Conquests of the lands]*. (M. Tavakol, Trans.). Tehran: Nashr-e Noghre.

Al-Heravi, S. M. (2004). *Tārikh-nāme-ye Herāt [The history of herat]*. Tehran: Asātir.

al-Maqdisi, A. M. (1982a). *Ahsan al-Taqāsim Fi Ma'refat al-Aqālim [The best divisions in the knowledge of the regions]*. Vol. 1. (A. N. Monzavi, Trans.). Tehran: Sherkat-e Moalefān va Motarjemān-e Irān.

al-Maqdisi, A. M. (1982b). *Ahsan al-Taqāsim Fi Ma'refat al-Aqālim [The best divisions in the knowledge of the regions]*. Vol. 2. (A. N. Monzavi, Trans.). Tehran: Sherkat-e Moalefān va Motarjemān-e Irān.

Amiri, H. (2001). Resaleh Tariq-e Qesmat-e Ab-e Qalb (in Persian). *Monthly Book of History and Geography, 44*(2001), 82–85.

Bal'ami, M. (1999). In M. Rowshan (Ed.), *Tarikh-e Bal'ami [Bal'ami's history]* (Vol. 1). Tehran: Soroosh.

Bartold, V. (1971). *Irrigation in Turkistan* (K. Keshavarz, Trans.). Tehran: Publication of Social Researches and Studies Institute.

Bartold, V. V. (1987). *Turkistān-nāme; Turkistān Dar Ahd-e Tahājom-e Moghol [Turkestan down to the Mongol invasion]*. (K. Keshavarz, Trans.). Tehran: Āgāh.

Beyhaqi, A. (1971). In A. A. Fayyāz (Ed.) *Tārikh-e Beyhaqi [Beyhaqi's history]*. Mashhahd: Dāneshgāh-e Mashhad.

Bichsel, C. (2016). Water and the (infra-)structure of political rule: a synthesis. *Water Alternatives, 9*(2), 356–372.

Boelens, R., Hoogestegerb, J., Swyngedouwc, E., Vosb, J., & Wester, P. (2016). Hydrosocial territories: a political ecology perspective. *Water International, 41*(1), 1–14.

Bosworth, C. E. (2007). The early ghaznavids. In R. N. Frye (Ed.), *The Cambridge history of Iran, Vol. 4; The period from the Arab invasion to the Saljuqs* (pp. 162–197). Cambridge: Cambridge University Press.

Briant, P. (2005). *Ta'mol Darbāre-ye Dowlat Dar Irān-e Bāstān; Dowlat-e Hakhamaneshi [Government in Ancient Persia; Achaemenid Government]*. In M. T. Boroomand (Ed. & Trans.), Jāme'-e Madani Az Didgāh-e Kārl Mārks [Civil Society from Karl Marx's View Point] (pp. 109–124). Tehran: Pāzi Tiger.

Budds, J., Linton, J., & McDonnell, R. (2014). The hydrosocial cycle. *Geoforum, 57*(2014), 167–169.

Butzer, K. W. (2012). Collapse, environment, and society. *Proceedings of the National Academy of Sciences of the United States of America, 109*(10), 3632–3639.

Chardin, J. (1966a). *Safar-nāme-ye Chardin [The Travels of Sir John Chardin]* (Vol. 3) (M. Abbasi, Trans.). Tehran: Amirkabir.

Chardin, J. (1966b). *The travels of Sir John Chardin (Persian translation)* (Vol. 4) (M. Abbasi, Trans.). Tehran: Amirkabir.

Christensen, A. E. (1989). *Irān Dar Zamān-e Sāsāniān [L'Iran sous les Sassanides]*. (R. Yasemi, Trans.). Tehran: Donyā-ye Ketāb.

Christensen, P. (2016). *The decline of Iranshahr; irrigation and environment in the Middle East 500 BC–AD 1500*. (S. Sampson, Trans.). I.B.Tauris & Co Ltd.

D'Allemagne, H. R. (1956). *Du Khorassan au pays des Backhtiaris; trois mois de voyage en Perse*. (M. A. Farrevashi, Trans.). Tehran: Amir-Kabir.

Danesh-pajooh, M. T. (1968). Dastoor al-Molook-e Mirzā Rafi'ā [The Kings' Orders]. *Dāneshkade-ye Adabiyāt Va Oloom-e Ensāni Dāneshgāh-e Tahrān, 65 & 66*, 298–322.

Dashti, M. (2005). Joghrāfiyā-ye Tārikhi-ye Shabake-ye Āb-hāye Beyn al-Nahreyn [Historical geography of the water resources network in mesopotamia]. *Tārikh Pajoohān, 1*(2005), 27–72.

Diener, A. C., & Hagen, J. (2012). *Borders: a very short introduction*. Oxford University Press.

Dinar, S. (2007). Water wars? Conflict, cooperation, and negotiation over transboundary water. In V. I. Grover (Ed.), *Water a source of conflict or cooperation*. USA: Science Publishers.

Dinawari, A. (1992). *Ketāb al-Akhbār al-Tiwāl [Book of general history]*. (M. Mahdavi Damghani, Trans.). Tehran: Nashr-e Ney.

Elman, Y. (2004). Up to the ears in horses' necks: on Sassanian agricultural policy and private eminent domain. *JSIJ, 3*(2004), 95–149.

Erfanian, M. (2014). *Barresi-ye Mafhoom-e Iqta' [A study on the concept of Iqta']*. Tehran: Pajooheshkadeh-ye Showrā-ye Negahbān.

Estakhri, A. E. (1994). *Masālik al-Mamālik [Routes of the Realms]*. (M. Tostari, Trans.), (I. Afshar, Ed.), Tehran: Bonyād-e Moghoofāt-e Mahmood Afshār Yazdi.

Faravashi, B. (1992). *Farhang-e Farsi Be Pahlavi [Persian Pahlavi dictionary]*. Tehran: Daneshgah-e Tehran.

Fouache, E., Cosandey, C., Wormser, P., Kervran, M., & Labbaf Khaniki, R. A. (2011). The river of Nishapur. *Studia Iranica, 40*, 99–119.

Foucault, M. (2002). *The archaeology of knowledge* (A. M. Sheridan Smith, Trans). London: Routledge.

Frye, R. N. (2006). The political history of Iran under the Sassanians. In *The Cambridge history of Iran* (Vol. 3, No. 1). Cambridge University Press.

Hakim al-Nishāburi, A. A. (1996). *Tārikh-e Nishābur [History of Nishapur]*. Tehran: Nashr-e Āgah.

Hamavī Baghdādī, Y. (2004). *Mu'jam al-buldān [Dictionary of Countries]* (Vol. 1) (A. Munzavī & A. Mihr-Parvar, Trans.). Tehran: Sāzmān-i Mīrāth-i Farhangī-yi Kishvar.

Harris, M. (1977). *Cannibals and kings; the origins of cultures*. Glasgow, UK: William Collins Sons & Co Ltd.

Hole, F. (2007). *The archaeology of Western Iran (Persian Translation)*. (Z. Baseti, Trans.). Tehran: Samt.

Hoyland, R. G. (2016). The founding of Basra and its early development. In P. Collins (Ed.), *Basra its history, culture and heritage* (pp. 49–52). British Institute for the Study of Iraq.

Hunt, R. C., Hunt, E., Ahmed, G. M., Bennett, J. W., Cleek, R. K., Coy, P. E. B., et al. (1976). Canal irrigation and local social organization. *Current Anthropology, 3*(17), 389–411.

Hunter, B. M. (2020). Silent and unseen: stewardship of water infrastructural heritage. In C. Hein (Ed.), *Adaptive strategies for water heritage: past, present and future* (pp. 21–39). Cham: Springer.

Ibn al-Athir, A. (1992). *al-Kāmil: Tārikh-e Bozorg-e Eslām Va Irān [al-Kāmil: great history of Islam and Iran]*. (A. Halat & A. Khalili, Trans.). Tehran: Moassese-ye Matboo'āti Elmi.

Ibn al-Athir, A. (1994a). *al-Kāmil fit-Tārīkh [The complete history] (in Arabic)* (Vol. 5). Beirut: Institute of Arabic History.

Ibn al-Athir, A. (1994b). *al-Kāmil fit-Tārīkh [The complete history] (in Arabic)* (Vol. 7). Beirut: Institute of Arabic History.

Ibn al-Fuwati, A. (2002). *Al-Havadis al-Jamea'; Events of the seventh century after Hijra*. (A. Ayati, Trans.). Tehran: Cultural Relics and Celebrities Association.

Ibn al-Jawzi, A. (1995). In Ata M. A. (Ed.), *al-Montazim fi Tarikh al-Moluk al-Omam [History of kings and nations] (in Arabic)* (Vol. 15). Beirut: Dar al-Kotob al-Elmiyah.

Ibn Battuta (1997). *Safarnāme-ye Ibn Battuta [Ibn Battuta's Travel Account]* (Vol. 1). (M. A. Movahed, Trans.). Tehran: Āgah.

Ibn Fondoq, (1982). *Tārikh-e Beyhagh [History of Beyhagh]*. Tehran: Ketāb-forroosh-ye Forooghi.

Ibn Hawqal (1987). *Ibn Hawqal's travel account (in Persian)*. (J. Shoar, Trans.). Tehran: Amir-Kabir.

Ibn Kathir, I. (1998). In Y. al-Baghaee & S. J. al-Attar (Eds.), *Al-Bidāya wa-n-Nihāya (in Arabic)* (Vol. 7). Beirut: Dar al-Fikr.

Ibn Khaldun, A. (1996). *Moghadame-ye Ibn Khaldun [Ibn Khaldun's introduction]* (Vol. 1) (M. Parvin Gonabadi, Trans.). Tehran: Enteshārāt-e Elmi Va Farhangi.

Ibn Khordadbeh, (1889). *Ketāb al-Masālik w'al- Mamālik [Book of roads and kingdoms]*. Beirut: Dār Sādir.

Ibn Rustah, A. (1986). *Book of precious records (Persian translation)* (H. Qarechanloo, Trans.). Tehran: Amirkabir.

Jacobsen, T. (1987). Mesopotamian religions. In M. Eliade M. (Ed.), *The encyclopedia of religion*. New York: Macmillan.

Jacobsen, T., & Adams, R. M. (1958). Salt and silt in Ancient Mesopotamian agriculture. *Science, New Series, 128*(3334), 1251–1258.

Jankowski, A. (2016). A reserve of freedom: Remarks on the time visualisation for the historical maps. In P. Collins (Ed.), *Basra its History, Culture and Heritage* (pp. 19–22). British Institute for the Study of Iraq.

Kaempfer, E. (1981). *Safar-nāme-ye Kaempfer [Kaempfer's travels account]* (K. Jahandari, Trans.). Tehran: Khārazmi.

Kamareyi, S. M. (2005). *Rooznāme-ye Khāterāt-e Seyyed Mohammad-e Kamareyi [Kamareyi's diary]* (Vol. 1). Tehran: Asātir.

Katib Baghdadi, Q. (1991). *al-Kharaj* (H. Qarchanloo, Trans.). Tehran: Alborz.

Kavandi, S., & Shad-del, T. (2013). Ta'sir-e Avāmel-e Joghrāfiyā-yi Bar Akhlāgh [The impact of geographical factors on morals]. *Akhlāgh-e Zist-mohiti [Environmental Ethics], 3*(9), 89–121.

Kennedy, H. (2011). The feeding of the five hundred thousand: cities and agriculture in early Islamic Mesopotamia. *Iraq, 73*(2011), 177–199.

Khazraee, S. E. (2008). Shadorvan bridge-dam and difficulties of historical studies on shooshtar (in Persian). *Golestan Honar Journal, 13*, 113–122.

Khosravi, K. (1969). Irrigation and rural society in Iran (in Persian). *Social Sciences Essays, 3* (1969), 48–56.

Kramer, S. N. (1972). *Sumerian mythology*. Philadelphia: University of Pennsylvania Press.

Kuiper, K. (2011). *Mesopotamia: the world's earliest civilization*. Britannica Educational Publishing.

Labbaf Khaneiki, M. (2019). *Territorial water cooperation in the Central Plateau of Iran*. Switzerland: Springer.

Labbaf Khaniki, M. (2012). Moghadame-yi Bar Joghrāphiya-ye Edāri-Siyāsi-ye Irān Dar Dowre-ye Sāsāni [Iran's political-administrative geography in the Sassanian period]. *Pāj, 1* (2), 116–131.

Lawrence, D, & Wilkinson, T. J. (2017). The Northern and Western Borderlands of the Sasanian empire: contextualising the Roman/Byzantine and Sasanian Frontier. In E. W. Sauer (Ed.), *Sasanian Persia between Rome and the steppes of Eurasia* (pp. 99–125). Edinburgh University Press.

Le Strange, G. (1900). *Baghdad during the Abbasid Caliphate*. Oxford: Clarendon Press.

Le Strange, G. (1905). *The lands of the Eastern Caliphate; Mesopotamia, Persia, and Central Asia from the Moslem conquest to the time of Timur*. New York: Barnes & Noble Inc.

Lees, S. H. (1989). On irrigation and the conflict myth. *Current Anthropology, 30*(3), 343–344.

Mahmoodian, S., & Qayoomi Bidhendi, M. (2012). Norm of canals; water management system in Safavid Isfahan (in Persian). *Architecture and Municipal Engineering Journal, 5*(10), 141–153.

Malayeri, M. M. (1996). *Iran's history and culture in transition from Sassanian to Islamic Era (in Persian)* (Vol. 2). Tehran: Toos Publisher.

Malayeri, M. (2000). *History and culture of Iran (in Persian)*. Tehran: Toos Publisher.

Markwart, J. (1994). *Ērānshahr nach der Geographie des Ps. Moses Xoranac'i*. (M. Mir-Ahmadi, Trans.). Tehran: Tahoori.

Mauch, C., & Zeller, T. (2008). *Rivers in history: perspectives on waterways in Europe and North America*. University of Pittsburgh Press.

Mays, L. W. (2008). A very brief history of hydraulic technology during antiquity. *Environmental Fluid Mechanics, 2008*(8), 471–484. https://doi.org/10.1007/s10652-008-9095-2.

Mccullogh, A. B. (1983). Currency conversion in British North America. *Archivaria, 16*, 83–94.

Meftah, E. (1992). Historical geography of Murghab–Merv and Merv Rood (in Persian). *Historic Researches Journal, 6&7*(1992), 71–132.

Millon, R., Hall, C., & Diaz, M. (1962). Conflict in the Modern Teotihuacan irrigation system. *Comparative Studies in Society and History, 4*(4), 494–524.

Mitchell, W. P. (1973). The hydraulic hypothesis: a reappraisal. *Current Anthropology, 14*(5), 532–534.

Mustawfi, H. (1983). *Nuzhat al-Qolub [Pleasure of the hearts]* (G. Le Strange, Ed.), Tehran: Donyā-ye Ketāb.

Nardo, D. (2007). *Ancient Mesopotamia*. Farmington Hills: Greenhaven Press.

Narshakhi, A. M. (1984). *Tārikh-e Bukhārā [The history of Bukhara]* (A. al-Ghabavi, Trans.). (M. T. Modarres Taghavi, Ed.), Tehran: Toos.

Nissen, H. J., & Heine, P. (2009). *From Mesopotamia to Iraq, a concise history*. USA: The University of Chicago Press.

No'pasand, S. M., & Isavi, H. (2013). Siyāsat-hāye Eghtesādi-ye Nezām-hāye Hokoomati-ye Sāmāni Va Ghaznavi [Economic policies of the Samanid and Ghaznavid Governmental systems]. *Pajoohesh-nāme-ye Tārikh-e Ejtemāyi Va Eghtesādi, 2*(2), 99–125.

Noelle-Karimi, C. (2014). *The pearl in its midst; herat and the mapping of Khurasan (15th–19th centuries)*. Vienna: Austrian Academy of Sciences Press.

Papoli Yazdi, M., & Labbaf Khaneiki, M. (2001). Measurement units for water division in traditional irrigation systems (in Persian). *Geographical Researches Journal, 49&50*(2001), 47–73.

Peet, R. (1985). Introduction to the life and thought of Karl Wittfogel. *Antipode, 17*(1), 3–21.

Pourshariati, P. (2008). *Decline and fall of the Sasanian Empire*. New York: I.B.Tauris & Co Ltd.

Qanzvini, M. T. V. (2004). *Tārikh-e Jahānārā-ye Abbāsi [History of Abbasids]*. Tehran: Pajooheshgāh-e Oloom-e Ensāni Va Motāleā't-e Farhangi.

Qazvini, Z. (1994). *Āthār al-Belād Va Akhbār al-Ebād [Monument of places and history of god's Bondsmen]*. Tehran: Amir-Kabir.

Qomi, H. M. (1982). History of Qom (Persian translation) (Hasan ibn Ali Abdolmalek Qomi, Trans.). J. Tehrani (Ed.). Tehran: Toos.

Radkau, J. (2008). *Nature and power; a global history of the environment* (T. Dunlap, Trans.). Cambridge University Press.

Rajabzadeh, H. (2005). Irrigation examined through documents of Qajar Iran. In N. Kondo (Ed.), *Persian documents, social history of Iran and Turan in the fifteenth to nineteenth centuries*. London: Routledge.

Ravandi, M. (2003). *Tārikh-e Ejtemāyi-ye Irān [Social history of Iran]* (Vol. 3). Tehran: Enteshārāt-e Negāh.

Rzoska, J. (1980). *Euphrates and Tigris Mesopotamian ecology and destiny*. The Netherlands: Dr. W. Junk bv Publishers.

Salehi, K., & Amir Shekari, F. (2013). Reasons and consequences of Ghaznavids' invasion of India. *Journal of Subcontinent Researches, 5*(15), 153–166.

Salim, M. N. (2018). Siyāsat-e Eghtesādi-ye Safaviyeh Va Ta'sir-e Ān Bar Soghoot-e Dowlat [Economic policies of the Safavids and their role in the Government's fall]. *Pajoohesh-hāye Oloom-e Tārikhi, 10*(2), 97–116.

Salimi Moayed, S. (2000). The Legend on Qanat Genesis in Shahdad Region (in Persian). *Proceedings of International Qanat Conference*, Vol. 1. Yazd: Yazd Regional Water Authority.

Simpson, S. J. (2016). Ard Maysan and the land behind Basra in late antiquity. In P. Collins (Ed.), *Basra its history, culture and heritage* (pp. 35–47). British Institute for the Study of Iraq.

Simpson, S. J. (2017). Sasanian cities: archaeological perspectives on the Urban economy and built environment of an Empire. In E. W. Sauer (Ed.), *Sasanian Persia between Rome and the steppes of Eurasia* (pp. 21–50). Edinburgh University Press.

Somervill, B. A. (2010). *Empires of ancient Mesopotamia*. New York: Chelsea House.

Souli, M. B. Y. (1935). *al-Owraq; Stories of ar-Radi bi'llah and al-Muttaqi (in Arabic)* (Vol. 1). Egypt: al-Sawi Publisher.

Strang, V. (2016). Infrastructural relations: water, political power and the rise of a new despotic regime. *Water Alternatives, 9*(2), 292–318.

Stroshane, T. (1988). Walden pond and the capitalist state. *Berkeley Planning Journal, 3*(2), 123–139. https://doi.org/10.5070/BP33213184.

Swyngedouw, E. (2004). *Social power and the urbanization of water*. England: Oxford University Press.

Tavernier, J. B. (1990). *Safar-nāme-ye Tavernier [Tavernier's travels account]* (A. T. Noori, Trans.). Tehran: Sanāyi.

Ur, J. (2014). Households and the emergence of cities in Ancient Mesopotamia. *Cambridge Archaeological Journal, 24*(02), 249–268.

Verhoeven, K. (1998). Geomorphological research in the Mesopotamian flood plain. In H. Gasche & M. Tanret (Eds.), *Changing watercourses in Babylonia; towards a reconstruction of the ancient environment in lower Mesopotamia* (Vol. 1, pp. 159–240). Belgium: University of Ghent.

Wilkinson, T. J. (1997). Environmental fluctuations, agricultural production and collapse: a view from Bronze age upper Mesopotamia. In H. Nüzhet Dalfes, G. Kukla, & H. Weiss (Eds.), *Third millennium BC climate change and old world collapse* (pp. 67–106). Berlin: Springer.

Wilkinson, T. J. (2003). *Archeological landscapes of the Near East*. USA: The University of Arizona Press.

Wilkinson, T. J., & Rayne, L. (2010). Hydraulic landscapes and imperial power in the Near East. *Water History, 2*(2010), 115–144.

Wilkinson, T. J., Boucharlat, R., Ertsen, M. W., Gillmore, G., Kennet, D., Magee, P., et al. (2012). From Human Niche construction to imperial power: long-term trends in Ancient Iranian Water systems. *Water History, 4*(2012), 155–176.

Wittfogel, K. (1981). *Oriental despotism: a comparative study of total power*. USA: Vintage.

Worster, D. (1985). *Rivers of empire: water, aridity, and the growth of the American west* (1st ed.). New York: Pantheon Books.

Wright, H. T. (1977). Recent research on the origin of the state. *Annual Review of Anthropology, 6* (1977), 379–397.

Yarshater, E. (2006). Iranian national history. In *The Cambridge history of Iran* (Vol. 3, No. 1). Cambridge University Press.

Yusefi, S. (2008). Siyasat-e Eghtesadi Dar Dowre-ye Nokhost-e Ghaznavi [Economic policies during the early period of Ghaznavid]. *Miskawayh, 2*(8), 253–268.

Yusefifar, S., & Azadeh, S. (2013). Manāseb-e 'Amid Va 'Āmel Dar Dowre-ye Saljooghi [The posts of 'Amid and 'Āmel in the period of Saljuqs]. *Ganjine-ye Asnād, 2*(90), 42–67.

Zahabi, S. M. (1998). In O. A. Tedmeri (Ed.), *Tarikh al-Islam wawafayat al-Mashahir wal'aelam (in Arabic)* (Vol. 31). Beirut: Dar al-Kitab al-Arabi.

Chapter 4
Hydro-social Cohesion

Abstract This chapter focuses on local communities on the Iranian plateau, taking a holistic look at the systematic relationships between them and their water resources in a geographical–historical context. In Iran, water has always been the most crucial production factor, exerting a great influence on socio-economic structures. However, the availability of water in the region has not been stable over time, and accordingly the local people have always had to evolve adaptation strategies through both technologies and social dynamics. This chapter examines the collective response of the local communities to their changing water resources, which occurs mostly through developing water management systems and water-based social mechanisms. The inhabitants could have enhanced their adaptation to their fluctuating water resources through social cohesion.This chapter tries to show how water affects the make-up of a local community through a variety of water-related social mechanisms, like cooperation, qanat maintenance, division of labor and waqf (charitable endowments). The chapter makes a distinction between two types of social cohesion, "inner-territorial cohesion" and "trans-territorial cohesion", which are both associated with the geographical peculiarities of the water resources of a specific region. Inner-territorial cohesion pertains to a social bond that forms among the residents within a particular territory mainly based on their common interest in terms of their water resources, whereas trans-territorial cohesion refers to a social bond between different territories or neighboring communities within the hinterland of a particular qanat sharing the same water resource.

4.1 Introduction

In this chapter, social cohesion is the core concept, where individuals find more benefit and less cost in cooperative actions by managing common resources (Bruhn 2009). Social cohesion deals with the concept of social capital (Madonsela 2017) that is needed to ensure the sustainability of a society (Atkinson 2019). A common goal like water management serves as "collective directional movement" (Wilson and Mansour 2020) that is among the vital factors constituting social cohesion

M. Labbaf Khaneiki, *Cultural Dynamics of Water in Iranian Civilization*,
https://doi.org/10.1007/978-3-030-58900-4_4

(Basati et al. 2019). And water management pertains to a set of social orders agreed upon by the members of a particular local community in order to: 1—maintain their common water resources, 2—preserve the relevant water catchment and physical water system, 3—divide water among the shareholders, 4—regulate water property, water market and conflict arbitration (Labbaf Khaneiki 2019).

Qanat water management has entailed cooperative activities strengthening social cohesion. The loss of qanats not only takes its toll on water resources but also undermines social cohesion, which is manifested in outmigration from rural areas. Qanat has given rise to a comprehensive "water culture" that contributes to social cohesion in the local communities where people could have survived several significant environmental changes, political upheavals and socio-economic transitions. Water culture is premised on interactions between hydraulic technologies and hydro-social cycles (Swyngedouw 2009; Boelens et al. 2016; Sanaan 2020).

The inhabitants of the Iranian highlands could survive several significant environmental changes by developing specific adaptation strategies,[1] manifested in sophisticated water technologies and water management forms. They could endure even harsher climatic conditions some 800 years ago with extreme water deficits (Sharifi et al. 2013), and they did not buckle under such environmental pressures (Labbaf Khaneiki 2014). Populations in the Iranian highlands evolved a resilient water management system based on, and contributing to, social cohesion in their communities. A key element of their success was the introduction and running of sophisticated water supply systems well adapted to the environmental conditions, the famous qanats[2] (Papoli Yazdi and Labbaf Khaneiki 2000; Salimi Moayed 2000; Semsar Yazdi and Labbaf Khaneiki 2017, 2019; English 1998).

There is a rich literature on qanats, traditional water supply, indigenous irrigation and water management systems in Iran (Ahmadi et al. 2010; Amiri 2001; Bartold 1971; Bazza 2006; Beaumont 1968, 1989; Enayatollah et al. 1971; Estaji and Raith 2016; Garoosi 2003; Hajian and Hajian 2013; Janebollahi 1990; Jomehpour 2009; Lambton 1939; Mahmoodian and Qayoomi Bidhendi 2012; Madani 2008; Manuel et al. 2018; Megdiche-Kharrat et al. 2018; Nash et al. 2012; Papoli Yazdi and Labbaf Khaneiki 1998; Remini et al. 2014; Balali et al. 2009; Saedloo 1997). Also, there are many studies on social issues of rural regions in the Iranian highlands (Azkia et al. 2011). However, I want to show how "water" contributed to social cohesion through a variety of water-related social mechanisms like water cooperation and irrigation system. One may classify "social cohesion in the Iranian villages" into two different types that are systematically interconnected: "inner-territorial cohesion" and "trans-territorial cohesion", which have something to do with the peculiarities of different water resources. Inner-territorial cohesion

[1]Under adaptation, I understand the collective ability to keep pace with the changing environmental possibilities through a set of measures, which are taken or the techniques, which are changeably used by the members of a particular local community.

[2]Qanat consists of some shaft wells interconnected by a subterranean tunnel that drains out the groundwater seepage and conveys it onto the earth surface using the height difference between the two ends of the tunnel.

pertains to a social bond that forms among the villagers within a particular territory mainly based on their common interest in terms of their water resources, whereas trans-territorial cohesion refers to a social bond between different territories or neighboring villages within the hinterland of a particular qanat sharing the same water resource.

Also, water has contributed to the social resilience against the changing environment in the Iranian highlands. Resilience is usually referred to as "the ability of a system to absorb disturbance and still retain its basic function and structure" (Walker and Salt 2006, p. 1), which can be translated in this book into the ability of a local community to survive environmental pressures through hydraulic innovations and water-related societal systems. This resilience mostly took place through such water-related strategies or technologies as fallow-cropping patterns, virtual water, qanat extension, water division systems, and so on.

Around 2500 years ago when a dry period neared its peak, the first qanats showed up and evolved in the area. Surface water resources started to shrink, and the population turned to groundwater which could be tapped by means of qanats, albeit wherever the environmental and topographical conditions allowed. Different archeological studies ever conducted on the history of qanats in various regions all give credence to the fact that the advent of qanats and the outbreak of a dry period were almost coeval. As an instance, the inscription of the Assyrian King Sargon, which glorifies his eighth battle in 714 BC, points out a civilization in the region of Urmia where a special irrigation system akin to qanat was in place at the time, according to the lines 202 until 204 (Laessoe 1951, pp. 21–32). In Bam (Kerman Province), at the bottom of a qanat well, Adle (2007) unearthed a container dating from 500 BC, which provides clues about the antiquity of the qanat. Also, around the wells of Qasabe qanat in eastern Iran some pottery fragments have been found, whose design and color imply the antiquity of 2200 years ago (Labbaf Khaneiki 1997, pp. 271–298). The French archeologist Boucharlat (2003, pp. 161–172) comes to the conclusion that qanats must have been invented in the Iron Age some 800 years BC, after reviewing all the studies ever conducted on the archeology of qanats. The communities once living on the banks of permanent rivers in the central plateau of Iran came to terms with the recurring droughts by turning to the technique of qanat, which caused a substantial change in their production system. This gave rise to some major modifications in the economic structures, and consequently brought about a transformation in social and cultural structures.

Technological strategies usually ended up in some societal changes, which were adopted to better cope with the water-related challenges. Catchment management, qanat maintenance, water division system, irrigation cycle, water endowment (waqf), water market, water rituals and territorial water cooperation are all among the social aspects of water, which had a crucial function in enhancing the communities' adaptation to their changing water resources. There are many subtle links between social cohesion and level of adaptation to environmental strains over the course of history, and water management systems could have contributed to social cohesion and resilience in the local communities. For example, the irrigation cycle was not only a simple queue for water, but it served as a social unit whose members were strongly bonded. They felt attached to the people who shared the same water cycle, coordinating with each other in terms of not only irrigation but also other social affairs, for example religious ceremonies, financial partnership and dispute settling.

In Iran, water has always been the most important production factor without which agriculture was almost impossible. Compared to the growing population, water resources were limited, and climatic changes aggravated the situation for agrarian communities. Some intricate management systems came into existence and evolved in the course of history in order to handle the scarce water resources effectively. Such water management systems exerted a great influence on the social order as a whole. The social bonds in a water management system could spread to the other social realms and then result in stronger social cohesion in the local communities.

4.2 Water and Inner-territorial Cohesion

In contrast to the general notion that water as a scarce vital resource in the arid regions tends to foment social tensions and competitions, in many parts of Iran water scarcity serves as the first domino of social cohesion. Water scarcity entails water rationing that takes place according to madār. Madār or irrigation cycle is the most basic concept in the Iranian traditional water division systems. Water for irrigation is owned by shares. In fact, the farmers take turns bringing water to their lands. For a particular shareholder, the interval between two irrigations means an irrigation cycle or rotation pattern of irrigation, usually of between 6 and 21 days' duration. Irrigation cycle provides the basis for other cooperative activities some of which have nothing to do with water supply. For example, in Kharanaq, Yazd province, the residential area was fortified with a wall that ran around the village. According to the elderly locals, 11 watchtowers were once embedded in the wall at regular intervals, a few of which are still standing. The farmers had to guard the village by being on the watch at the watchtowers lest the burglars did not break into their houses and steal their belongings. The shift system was based on their 11-day irrigation cycle in order to determine who should have been on watch every day at each of those 11 watchtowers (Fig. 4.1).

All the shareholders abide by the qanat water division system, and everybody takes their shares according to an irrigation cycle, which allows them to irrigate their lands only at certain times. However, sometimes all the shareholders have to change their irrigation cycle to re-adapt to the diminishing water resources. Fluctuations in the flow of a qanat can drive the shareholders to modify their irrigation cycle unanimously. If the discharge of a qanat decreases due to a drought, the water shares no longer suffice to irrigate the existing lands. My field study in the village of Qasem Abad, Yazd province, shows how irrigation cycle serves as a ground for social cohesion that goes beyond all ethnic and religious differences. The village of Qasem Abad nestles on the outskirts of the city of Yazd, where a

Fig. 4.1 Kharanaq's 11 watchtowers whose shift system was taken from an 11-day irrigation cycle

qanat is running as the sole water source for irrigation. This qanat dates back to some 200 years ago, named after a famous charitable merchant Haj Abol-Qasem Rashti. This qanat is 23.35 km long in sum including its main tunnel and its side branches, which once enjoyed a discharge of 40 L per second in 1998 (Yazd Regional Water Authority 1999).

Traditionally, the water of Qasem Abad qanat is shared out between Zoroastrians and Muslims on an irrigation cycle of 12 days. Each day of the irrigation cycle has a special name. At present water is being divided among the farms and orchards based on an accurate management system left from the past. Water is measured and rationed based on a time unit called jorreh after the name of a traditional water clock that was once used to calculate the time of irrigation. The time unit is still called jorreh, though the water clock itself was made redundant by the modern wristwatch. A jorreh equals 11 min. Hence, a person who holds one jorreh has the right to irrigate his/her land with the qanat water for 11 min. In sum, the qanat water is divided into 1560 jorreh of which 340 jorreh belong to Zoroastrian farmers, distributed over a 12-day irrigation cycle. Each day is divided into eight subdivisions and each subdivision contains 16 jorreh. Therefore, each day is expected to contain 128 jorreh. However, in practice each day is made up of 130 jorreh of which 128 jorreh belong to the shareholders and 2 additional jorreh go to the water chief—who is in charge of water division—as his wage. Thus, the 12-day irrigation cycle is constituted of 1560 jorreh. Each day of the irrigation cycle is named after someone who owns the most water shares on the same day, as follows: Foroodi, Askari, Khoramshahi, Qolami, Maryam Abadi, Ali Akbari, Haji Ali, Mohammadi, Sorkh Abi, Ramezani, Shahr Bamesi, and Qolam Rezayi.

The 12-day irrigation cycle is regarded as the standard cycle suitable for wheat as their routine cultivation. In other words, the 12-day cycle fits in with the water demand of wheat cultivation, whereas the other crops with higher water demand should be irrigated at closer intervals and on a shorter irrigation cycle. For example, in the case of orchards, the irrigation cycle varies from 6 to 12 days with the type of fruit trees. Such crops as turnip and beet require a 6-day irrigation cycle, and the crops like cucumber, okra and tomato need a cycle as short as 3 days.

Therefore, the water chief has to deal with different groups of shareholders with different cropping patterns and thus different water demands. Some of them want to irrigate once every 12 days, whereas some prefer a 6-day cycle and some a 3-day cycle. Moreover, a particular shareholder may get a part of his water share on a 12-day cycle, and the other part on a 6 or 3-day cycle. For example, when someone owns 8 jorreh, he can receive either the entire 8 jorreh at once on a 12-day cycle or 4 jorreh twice on two successive 6-day cycles.

The first solution to cope with such a stunning complication was to clump together the crops with similar water demands. In other words, the water chief advises the farmers to grow the crops with similar water demands in a particular cultivation area, so that water can be transferred more easily with minimum waste and on the same irrigation cycle. Clustering the crops with the same water demand makes it possible for the farmers to team up at different stages of their cultivation from plowing to sowing to irrigating to reaping, which benefits everyone alike. All the neighbors who decide to grow wheat for example cooperate to plow and sow everyone's lands and then jointly reap the crops.

The second solution is to break down each water share and distribute it over the irrigation cycle, which is impossible without cooperation. For example, on the first day of the irrigation cycle, which is called Foroodi day, the farmers are supposed to irrigate once every 12 days, but those who want to get their water once every 6 days have to cooperate with the shareholders of the seventh day or Haji Ali day. In fact, the farmers of Foroodi day receive half of their water share and wait for 6 more days to get the rest on Haji Ali day. This practice necessitates cooperation from the farmers of Haji Ali day, who accept to get half of their water share on Foroodi day in exchange. To do so, the farmers of the 2 days should coordinate their cropping patterns to ensure a similar water demand. On the other hand, day or night irrigation is another issue that should be taken into consideration while exchanging their places on the irrigation cycle. Someone on the first day (Foroodi day) who has the right to irrigate in the daytime prefers to swap his place with someone who irrigates in the daytime too, though the day and night irrigations alternate over the next cycles. Thus, a person irrigating at night would be allowed to irrigate in the daytime during the next cycle. Foroodi and Haji Ali days are 6 days apart, and cooperation between their shareholders works only when they all want to irrigate their crops on a 6-day cycle.

As for the crops that should be irrigated on a 3-day cycle, the farmers of Foroodi day need to cooperate with those of Qolami day 3 days ahead. Therefore they receive part of their water shares 3 days later on Qolami day in order to water the crops in need of more frequent irrigation. In return, the farmers of Qolami day receive the same water share 3 days later on Haji Ali day, and then the farmers of Haji Ali day shift the same amount of water shares to 3 days ahead on the cycle, by handing over the same water shares to the farmers of Qolami day and getting the same share on Ramezani day instead. The farmers of Ramezani day get back from Foroodi day what they have given to Haji Ali day. This way they complete the loop of cooperation. The same order applies to the other days of the irrigation cycle as seen in Fig. 4.2.

In the village of Qasem Abad, water is the first ring of the production chain, without which agriculture and life are both unattainable. A self-centered person on the irrigation cycle can bog down the entire production system, if he/she refuses to stay in harmony with other farmers. The irrigation cycle—as the cornerstone of their production system—makes it possible for all inhabitants to regulate their relationship with each other on the one hand and with the limited water resource on the other hand. But this cycle should not remain static, since the water resource correlates with the annual precipitation whose fluctuation is inevitable. Hence, all farmers have to work in tandem in order to keep their irrigation cycle in balance with the changing water resource. When water dwindles, all farmers pull together to change the irrigation cycle in order to minimize its impact on individuals by sharing out loss of earnings among the community. And when water increases, they cooperate to change the cycle again to make the most of the available water. In this example, we see how all persons work together like cogs in the irrigation system where social attachment becomes the key to survival in such a harsh environment. The sustainable utilization of the scarce water resource entails a high level of cooperation, and accordingly social cohesion that does not remain limited to

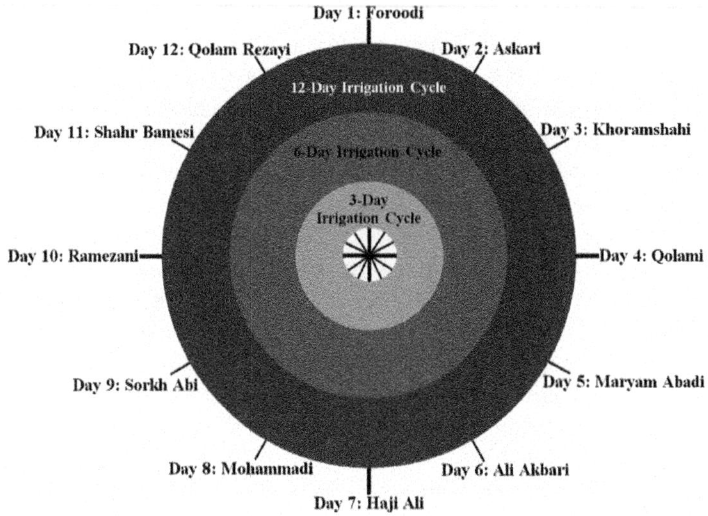

Fig. 4.2 Combination of different irrigation cycles in the qanat of Qasem Abad, based on water cooperation (Foroodi day is assumed as a benchmark)

irrigation system, but it is systematically extended to the other realms of social life. All villagers know that any falling-out (even irrelevant to irrigation) with a fellow farmer may catch up with them on the irrigation cycle, where they are highly dependent on cooperation. Hence, they try to compromise on such disputes as far as possible, and even those who are richer do penance by giving a part of their revenues to charity, in order to protect their cooperation-based irrigation system against any potential social disorder.

Such charitable endowments are called waqf that is believed to belong to God. The donor is exalted as a righteous person by hyping his deed as a pious endeavor motivated by Quran the Muslims' holy book (Baqutayan et al. 2018, p. 149) or Avesta the Zoroastrian scriptures (Gheybi 2011, p. 23). However, in fact the donor consciously or unconsciously wishes to avoid entering the troubled waters of social class conflict. The irrigation system is highly vulnerable to any antagonistic class struggle, taking into account that the qanat water has not been shared out among the villagers equally. Some villagers have the right to irrigate only for a few minutes, whereas some can have the flow on their farmlands for half a day or even more. More water results in vaster cultivated area and then higher income, giving rise to class inequality that should be kept at bay through such socio-economic strategies as waqf. Otherwise, those with less water shares may shirk their duty as cogs in the irrigation system, and as a result the entire economy would collapse. Hence, the rich allocate a small percentage of their revenues to the public by charitably building and maintaining such amenities as water reservoir, irrigation house, and so on, in order to insure their economic status against any threat induced by class inequality. In our modern capitalist society, this task has been handed over to the insurance companies to a large extent, but in the past waqf proved effective to keep a balance between social classes.

On the farmlands of Qasem Abad, a number of "irrigation houses" are visible, spreading throughout the landscape. An irrigation house is called gombeh in the local dialect, which is a mud brick building with the dimensions 4 by 4 by 2 m. Its front porch is covered by an adobe arch, where a small entrance opens into the interior space. Gombeh accommodates the farmers while waiting for their turns to irrigate their lands that are too far from their homes in the village to commute every day. The farmers rest on the front porch usually in spring and fall, whereas they sleep on the rooftop at summer nights, and take shelter inside the building in winter. Inside each gombeh, they operate a water clock to calculate everyone's time of irrigation. Each gombeh is named after its founder like Key-khosrow, Nāmdār, Sorkhāb, Hakim, Tal-varji, and so on, all of whom are among the rich villagers. Whoever builds a gombeh for public profit is responsible for its maintenance cost as well, and his sons inherit this charitable duty after his death. This way, the wealth is transformed from "an unfair access to more resources" to "an opportunity bestowed by the upper class" (Fig. 4.3).

Therefore, such socio-economic structures can buffer the effects of social inequality. An orderly irrigation system in turn leads to higher social cohesion in the village. Social cohesion insures them against revenue loss because of illness, travel or family problems. Its mechanism is simple but very effective. They need their

Fig. 4.3 Gombeh, where water shares are calculated and the farmers rest while waiting for their turns to irrigate

fellow farmers to be in sync while dividing water on a dynamic irrigation cycle. They show much prudence in treating each other on every occasion lest anyone retaliate by refusing to cooperate at the time of irrigation. For example, if they get into a scuffle with a neighbor, they can easily risk their cultivation when they would be in need of irrigation at shorter intervals but the same neighbor would refuse to exchange his water share on the irrigation cycle. Thus, such "irrigational bond" is translated into social cohesion that insures the farmers against other threats that are seemingly irrelevant to irrigation. In the village of Qasem Abad, if a farmer falls ill or goes on a necessary trip, the agricultural calendar does not wait for him to recover or return. All agricultural chores like plowing, sowing, irrigating and reaping should be done respectively on time. In such cases, the other farmers undertake to do all those chores on his behalf to secure his harvest. Otherwise, a simple illness can leave his land uncultivated and put him through bankruptcy and even starvation.

In this village, cooperation is not limited to when a farmer is in such troubles as crippling illnesses or imperative travels. Many of the agricultural chores demand a high level of cooperation that is engendered by the irrigation cycle in the first place. For example, plowing by shovel is a laborious job that cannot be done by a farmer alone. An individual can shovel soil barely more than 300 m^2 a day, which is really slow. It is possible to work the land with cows and plough, but only in the case of wheat and barley cultivation. For the other crops, plowing should be to a depth of more than 15 cm, which is not possible with an animal-driven plough. Hence, the farmers rely on each other's cooperation to plow their lands on time according to their agricultural calendar. Their social cohesion facilitates such cooperative activities that can enhance the efficiency of their production system. If a farmer insists on shoveling alone, he would lag behind the farming schedule and his toil would never come to fruition, which would lead to bankruptcy and even starvation again. For shoveling, a perfect team is made up of four farmers two of whom are left-handed and two right-handed. They work shoulder to shoulder in a row, such that two farmers move earth to their right side and two farmers to their left side, and this way a furrow appears in between, which is later necessary for irrigating the crops. They push their shovels into the soil simultaneously and fully in sync, to the rhythm of a traditional song. When they finish with one's land, they immediately shovel other lands one after another. Thus, the four farmers plow all four lands together quicker than when each works his own land alone, because they have to keep pace with the group having less time to rest, and the irrigation furrows are automatically built, which otherwise have to be devised later between the farming rows. Social cohesion proves to be a context in which economy (agriculture), biology (handedness) and art (traditional song) are juxtaposed in order to enhance the inhabitants' adaptation to the environmental conditions working behind their agricultural calendar (Fig. 4.4).

In the village of Qasem Abad, irrigation system has become the basis for social cohesion to the extent that even religious differences have dimmed. Majority of the village's inhabitants are Zoroastrians, interspersed with patches populated by some groups of Muslims and Baha'is across the village. In the village, a local dialect is spoken called Dari or Gabri, which is specific to Zoroastrians all over the region (Vahman and Asatrian 2002, p. 18). But the adherents of other faiths commonly use the same language that can smooth the way for a better communication and deeper understanding among all inhabitants. In a nutshell, the farmers resort to the irrigation cycle as an adaptation strategy, which entails a great deal of cooperation. Such a strategic cooperation helps strengthen their social cohesion in general to the extent that many cultural differences between the village's social groups recede, fostering kind of linguistic integration.

Fig. 4.4 Left-handed and right-handed farmers shoveling their lands in harmony (by Kiana Labbaf Khaniki)

4.3 Water and Trans-territorial Cohesion

Trans-territorial cohesion is well embodied in a groundwater reserve shared by different territories along a qanat. This situation is associated with the qanat's length, which runs across a vast arable area. The first question that may flash through your mind is how different communities can peacefully share a long qanat without any top-down authority to regulate their relationship. I try to answer this seemingly simple but in fact complicated question.

Ostrom's theory on common property can help us better understand the socio-economic structure behind a cooperative water management based on qanat (Ostrom 1990), answering the question why the relationship between the qanat upstream and downstream does not reflect the logic of prisoner dilemma in the game theory. The advocates of the game theory believe that such cooperation can be explained through a tit-for-tat relationship that the individuals establish between themselves according to a spontaneous cost-profit calculation. This relationship that can lead to a cycle of cooperation has been formulated in the game theory (Axelrod 1984).

However, the genuine cooperative interactions between the qanat upstream and downstream do not always live up to the logic of the prisoner dilemma. According to the prisoner dilemma, an over-exploitation in the upstream would be responded by a similar action in the downstream and their relationship would spill into a vicious cycle of retaliation until their shared resource would collapse. Hardin calls this situation "the tragedy of the commons" resulting from unlimited demands for a limited common resource. Under such condition, everyone in a particular community tries to gain more share of a common resource in the knowledge that over-exploitation would put pressure on the resource and bring about a long-term cost, but he would be incentivized by the fact that the cost is for everyone and the benefit is only for him. The common resources tend to enter the tragedy of the commons unless some systematic factors would curtail the over-exploitation and discourage the rapacious individuals (Hardin 1968). However, in realty there are other factors that rule the game, the factors that are embodied in the aforementioned socio-economic structure. This structure bestows sustainability and resilience upon the relationship between the qanat upstream and downstream, ensuring a fair and peaceful exploitation of their shared water resource.

Therefore, we should take a further step toward a more convincing theory to better understand the concept of water cooperation in a particular qanat. We should not confuse common pool resources with open source properties, which are believed to have been pervasive in the primitive societies within their communal territory where everyone—from the same society—was equally entitled to the available vital resources almost with no regulation or limitation (Fennell 2011, p. 12). Groundwater as a common pool resource is controlled by a nexus of social and economic processes, which regulates the individuals' access based on their social roles, positions, and so on. Given that the groundwater resource belongs to the whole society rather than to a few individuals, the utilization of groundwater remains within the bounds of the social circumstances on the one hand and the environmental conditions on the other hand. Thus, treating the groundwater as a common pool resource could have ensured its sustainability to the long-term benefit of the beneficiary society in the course of history.

The hinterland of a particular qanat can be understood through the model of "common benefit" according to which an intricate set of socio-economic factors knits together the interests of the basin upstream and downstream. One of those factors is the spatial extension of qanat, which may transcend different territorial borders. Unlike wells, qanat is not just a point on the landscape, but its linear structure begins at a mountain base and stretches down to the less elevated lands, spanning tens of kilometers. For example, in the case of Hasan Abad qanat, in Yazd province, the qanat's mother well which is the last and deepest well is situated some 40 km away from the qanat's exit point in the village of Hasan Abad. The water of this qanat is shared out between the town of Mehriz, the village of Dehno and the village of Hasan Abad. Annual maintenance of such a long gallery along with hundreds of shaft wells is quite unaffordable for the farmers of Hasan Abad alone, in the qanat downstream, if there would be no one else to contribute. Moreover, the recharge area of the qanat's aquifer is located in the upstream where a vast area of land has been allocated to the aquifer recharge. Thus, the downstream needs the upstream for their cooperation and the upstream can have a share of the qanat's water in return. Also, it is almost impossible for the upstream to take care of the qanat without the cooperation of the downstream, so the upstream can ensure the qanat's sustainability by observing the right of the downstream.

Moreover, in the cultural landscape of Hasan Abad qanat, there are other factors that accentuate the common benefit. Some of those factors used to play a crucial role in creating the common benefit, but they have lost their traditional function today. As an instance, we can mention several watermills that were built along the qanat gallery all the way from Mehriz to Hasan Abad. In the upstream, the farmers managed to build some watermills whose revenue could cover a part of the qanat maintenance cost, to the benefit of the downstream as well. The same relationship was in place regarding a land that was entrusted to the ditch watchman, where he could cultivate in return for his service. The land was located in Mehriz (qanat upstream), but the watchman's service was rendered to the downstream.

The qanat water was not used only for irrigation, but its high quality also lent itself to supplying drinking water to the three territories. The water reservoirs in the three territories were filled up with the water of Hasan Abad qanat, and the maintenance cost of the same reservoirs were paid through some charity water shares of the same qanat, which had been donated to public water supply according to the tradition of waqf. The charity water shares were rented out and their revenue was spent on the repair and maintenance of the water reservoir. The important role of Hasan Abad qanat in supplying drinking water galvanized all the beneficiaries into closer cooperation in order to maintain the qanat and its traditional functions.

Such complicated interactions underlie a special socio-economic structure that in turn gives rise to territorial water cooperation.

Cooperation in the qanat's maintenance and preservation can equally benefit the three territories, and this common benefit creates a vital triangle including water, food and energy. As mentioned, the qanat of Hasan Abad used to supply safe drinking water to the three territories. Also, the water of this qanat was used to irrigate the farmlands that provided the three territories with food. Moreover, the qanat directly and indirectly provided the energy demanded by the three territories. The watermills working with the same qanat generated the kinetic energy required for flour production. But thermal energy was of more importance, which was highly demanded for cooking and heating their homes during winter. The three territories enjoyed many orchards all irrigated by the same qanat, where the trees had to be pruned once a year in order to keep bearing fruit. As a result, a considerable amount of firewood was produced, which could fulfill the three territories' demand for energy. Providing water, food and energy was a common benefit whose permanence depended largely on such cooperation between the three territories. The model of common benefit simply hinges on the fact that cooperation between the three territories can ensure the supply of the vital triangle of water, food and energy. Otherwise, all territories would sustain a loss alike.

Although in the case of Hasan Abad qanat, territorial water cooperation is still in force and is not replaced by a competitive confrontation, the main factors that underlie the common benefit are changing one after another. Those factors are like some strands tied together to form a rope, but the strands are being snapped one after another, and just a few strands are now left undamaged. The watermills have lost their traditional function and have sunk into oblivion. The water reservoirs have been made redundant by the modern tap water network, and no longer contribute to the region's common benefit. Nonetheless, to date the agricultural activities have survived, and are still dependent on the qanat as a common water source which still knits together everyone's benefits. However, agriculture will not be immortal under the pressure of climate change, urban sprawl and water mismanagement. If the modern development would not be reconciled with the environmental capacities, the last strand would be snapped soon and the social cohesion based on water cooperation would plunge into the abyss of annihilation.

Trans-territorial cohesion does not revolve only around qanat's water cooperation, but it may be rooted in other forms of territorial water cooperation. As an example, I wish to touch upon the Abarkooh basin in central Iran, where the upstream is occupied by nomads with an economy mostly based on animal husbandry, but its downstream is populated by agrarian communities. Abarkooh exemplifies the conditions that systematically lead to territorial cooperation on shared water resources, the conditions whose disproportional change later turned cooperation into competition over the same water resources.

The basin upstream contains the town of Abarkooh in Yazd province and its related residential centers, and the basin downstream includes the towns of Abadeh and Eqlid in Fars province along with their villages. From about 300 to 50 years ago, the basin upstream and downstream used to live in harmony and in a systematic balance which can be called a period of livelihood differentiation. Livelihood differentiation is the settling of various livelihoods in different geographical spaces that best match the economic capacities and ecological requirements of each space. Livelihood differentiation can be regarded as a kind of adaptation strategy to the climatic and environmental conditions. Nomadic livelihood based on animal husbandry settled in the basin upstream, whereas agricultural livelihood occupied the basin downstream and flourished there. Thus, the vital resources (water and soil) were fairly divided among them, and their different niches and needs reduced the risk of any conflict between them over the region's limited resources. Nomadic livelihood was deeply dependent on pastures and handicrafts, whereas agricultural livelihood relied on water and soil. Therefore, the water could be released for the downstream where arable soil was abundant and a lucrative agriculture was possible. In return, their agricultural surplus could satisfy the need of people in the basin upstream for such products. Also, the basin downstream could profit from the dairy products and handicrafts produced in the upstream. Agriculture in the fertile soil found in the basin downstream was quite rational, and animal husbandry and producing handicrafts in the upstream made economic sense.

However, over the past five decades, a nexus of complicated human and natural factors unbalanced the systematic interactions between the basin upstream and downstream and pushed them into a longstanding competition over the limited water resources, which brought about nothing but over-exploitation of groundwater and escalating water crisis. In the Abarkooh basin, the groundwater depletion has resulted in an extensive land subsidence that is marked by 33 sinkholes across the plain. Abarkooh holds this lesson for us that the economies in the upstream and downstream should be complementary, like different pieces of a jigsaw puzzle which can fit together to show the beautiful picture of sustainability. For example, in the upstream, instead of cancerous growth of apricot orchards that are insatiably sucking up the shared aquifer, food processing industries could have been established and developed, because these industries would have been dependent on the downstream agricultural products as the last ring of a complementary production chain. Moreover, such industries would not have consumed a considerable amount of water, and most of the shared water could have been released for the downstream cultivation that could have provided those industries with their needed raw materials. All those can happen in the paradigm of sustainable development where systematic ecological interactions, environmental ethics and social justice go hand-in-hand.

There are many drivers at play in the competitive confrontation between the basin upstream and downstream, out of which governmental intervention and political decisions are of great importance. Also, one can find other countries where —behind the scenes—political decisions undermine water cooperation between the local communities. For example, in Vietnam, Bolivia and Zambia, governmental intervention sometimes contributes to more competitive confrontation between the water stakeholders. The government-backed water organizations pave the way for the wealthy households to exert more influence on water management, and the poor households shy away from pushing for their actual water rights because of the rich's hegemony over other socio-economic aspects of their lives. Such water organizations can widen the gap between the social classes and rather destabilize water governance in the long term (Funder et al. 2012). In Bolivia, the governmental inter-basin water transfer projects change the stakeholders' perception of their hydro-social territories, and such projects turn into an arena where stakeholders struggle to get their imagined hydro-social territories accepted and dominant (Rocha López et al. 2019: 282). Water resources that have been traditionally shared between local communities are re-allocated based on power relations (Swyngedouw 2004). Re-allocation of water resources benefits some social groups in favor of power relations, whereas it wreaks havoc on stability or coherence of other social groups or environments (Swyngedouw 2009, p. 57). Power relations contribute to a wider social disparity in terms of access to water resources (Moncrieffe 2004), and such disparity impedes the sustainable implementation of integrated water resources management which requires stakeholders to jointly manage their water resources, rather than fight over them (Perret et al. 2006, p. 149).

By investigating water cooperation in Abarkooh basin, we can conclude that development as an absolute phenomenon based on "power relations" and out of the ecological and social contexts bears no value. In every natural system from a tiny cell to a huge ecosystem, development can ensure the system sustainability on condition that the four following vital processes would remain in balance: material and energy input into the system, material and energy output from the system, organic relationship between the system components, and organic relationship between the system as a whole and its outside environment.

If a system keeps growing beyond its physical intrinsic limitations to the extent that the above-mentioned processes cannot function properly, the demise of the system would be inevitable, because the size of system grows out of proportion to its possible inputs and outputs, and the extraordinary size leads to a disconnection between the system components and also impedes the organic relationship between the system and its surroundings. Human societies are socio-economic systems whose disproportional development can hamper the aforesaid vital processes, leading to their collapse. Human history is awash with many stories about the famous civilizations whose disproportional development dragged them into annihilation (Diamond 2005). Such great civilizations lost their organic relationship with their natural environments due to their runaway development that eventually burst into their collapse. Abarkooh basin is not exempted from the same destiny if their sustainable development would not get back on track. The rapid development of agriculture in the basin upstream and downstream simultaneously without taking into account their different ecological capacities would aggravate the water crisis and bring their economic systems to the brink of annihilation.

I do not claim that the Abarkooh model can be extrapolated to every region where the upstream is occupied by nomads and the downstream is populated by agrarian sedentary societies, but it is not deniable that a similar model may be found to some extent in other regions even on a much larger scale. As an instance, when we follow the rivers Euphrates and Tigris up to their origin, we reach the Taurus

Mountains in eastern Anatolia in Turkey where nomadism has long stood out as the prevailing life style. However, in the downstream the Mesopotamian civilizations have long been anchored in an agrarian economy based on irrigation. In the rivers upstream, the mass influx of Turkish pastoralists from the end of the eleventh century onwards could drastically change the demographic composition of the region in favor of nomadism (De Planhol 2004, p. 525). However, later the Turkish nomads moved westward and tended to become more sedentary, creating a vacuum in eastern Anatolia, which was gradually filled by Kurdish pastoralists. The Ottoman Empire encouraged the Kurdish pastoralists to replace the Turkish nomads whose absence left their eastern frontiers defenseless against the Persian invasions (Ibid, p. 528). The nomadic population reached its peak from the late sixteenth century to the early seventeenth century, so that the pastoralists overwhelmingly outnumbered the sedentary people by more than four to one (Dede 2011, p. 41). However, the story changed when Ottoman administration embarked on the Tanzimat reforms in the nineteenth century, according to which a sedentarization program was imposed on the nomadic societies in eastern Anatolia (Ibid, p. 26). From then on, sedentarization and accordingly agriculture were on the rise in the upstream, leading to an increasing demand for irrigation water. Turkey managed to build 22 dams on these rivers to keep pace with the growing demand for water in the upstream regions where agriculture had become the hub of the economy. The downstream regions in Syria and Iraq were almost deprived of their historical shares of these two rivers, and their agrarian societies eventually succumbed to a set of complicated socio-economic problems mostly brought about by water scarcity. Deteriorating social cohesion, the rise of extremism and longstanding political turmoil in the downstream countries (Syria and Iraq) seem to have something to do with the sedentarization and massive agricultural development in the upstream over the past century to the detriment of traditional socio-economic structures in the downstream.

4.4 Conclusion

About 300 years ago, Hobbes said that before the advent of governments, the world used to teem with selfish peoples who never succumbed to cooperation, and turned life into the short moments of emptiness, loneliness and ruthlessness. According to him, cooperation is fully unattainable in the absence of an exogenous force like government that can put all individuals together, regulate their common resources and navigate them toward a common goal (Hobbes 1962: 100). However, recent archeological discoveries have taught us that cooperation was a pervasive social act even among the prehistoric communities long before the genesis of any kind of government (Carballo 2013, p. 20), and it was motivated by combinations of mechanisms depending on social context (Ibid, p. 5). One of those mechanisms was environmental condition that coerced the individuals into cooperation, in combination with other mechanisms like production system. If Hobbes replaced "government" with "environment" in his argument, he could probably attain more advocates these days. In Iran, environmental conditions put restrictions on available water resources, such that an elaborate water division system is developed as a collective solution to water scarcity. The water division system motivates and sustains cooperation at two different levels: within the local communities and between them. More cooperation paves the way for a more cohesive society that proves more adaptable to such a harsh environment.

In the Iranian plateau, inner-territorial cohesion hinges on irrigation system. An irrigation cycle requires a high level of cooperation that helps all water shareholders stay adaptable to their fluctuating qanat flow. When the qanat flow dwindles, the shareholders unanimously change the irrigation cycle to keep a balance between the amount of available water and the crop water need. A non-cooperative person can break the entire chain of cooperation on the irrigation cycle, which results in a malfunction in the village's production system. Hence, everybody tries to avoid any conflict even ones irrelevant to irrigation, in order to get away from any retaliatory reaction targeted at the irrigation cycle. The cost of any potential conflict outweighs its benefit for either side. Thus, almost all social behaviors are systematically oriented toward convergence and consensus, which underlie an efficient irrigation cycle. The tradition of waqf (charitable endowment) serves the same function, where the rich donates a portion of their revenues in order to dampen down the class tensions that can potentially beset the vulnerable irrigation system.

Common water resources may contribute to the trans-territorial cohesion as well, through a network of socio-economic interactions between the upstream and downstream areas. This chapter examined the shared water resources and their role in establishing trans-territorial cohesion, by making a comparison between two regions in the province of Yazd, Iran. In the first region, the common water resource still serves as a basis for cooperation and social cohesion, whereas in the second region over the last four decades territorial water cooperation had begun waning and social tensions had taken the place of cooperation.

Two different conceptual models can be suggested to describe the situation of those regions in relation to their common water resources; the model of "common benefit" and the model of "livelihood differentiation". The model of common benefit is premised on an intricate set of socio-economic factors that knit together the interests of the basin upstream and downstream in the light of "water-food-energy nexus". This model explains how several territories with similar economic structures could have developed a cooperative water management system that ensures sustainable water equality. As an instance, the water of Hasan Abad qanat is shared out between the town of Mehriz, the village of Dehno and the village of Hasan Abad on the outskirts of Yazd, along its 40 km length. One of the most important properties of this qanat is its territorial cooperation based on a lax territorial behavior, which is associated with the qanat's dynamics. Dormant territorial behavior can pave the way for a high sense of cooperation and social capital, which are all rooted in the peculiarities of qanat.

By livelihood differentiation, I mean a geographical disparity between the basin upstream and downstream, which provides different possibilities for rural communities, and different possibilities in turn give rise to diverse economic systems with different water demands. As an example, this chapter examined the situation of the Abarkooh basin in central Iran, where the upstream is occupied by nomads with an economy mostly based on animal husbandry, but its downstream is populated by agrarian communities. This situation systematically led to territorial cooperation on shared water resource, though some asymmetrical changes later turned cooperation into competition over the same water resource. Traditionally, the different geographical conditions across the basin drove each community to occupy a particular niche most suited to their livelihood. The upstream enjoys porous and highly permeable soil that does not favor irrigated farming, plus good pastures and abundance of water. This condition attracts the people who live off livestock and animal husbandry. On the other hand, the downstream lacks that amount of water and good pastures, but it has a fertile soil ideal for irrigation. This condition is to the farmers' liking, of course, if the upstream people let the water flow down. Water demands in upstream and downstream economies used to be in harmony, because of their structural differences matching their diverse geographical backgrounds. Their economic interdependence sustained livelihood differentiation model for centuries, where the basin upstream bartered their livestock products for the agricultural products of the basin downstream and vice versa. Hence, geographical diversity can lead to an environmental justice in terms of shared water resources, if our economies would be configured in compliance with geographical possibilities.

References

Adle, S. (2007). *Qanāts: an archaeological perspective*. International Training Course on Qanats, Yazd, IR Iran, 1–4 July 2007.

Ahmadi, H., Nazari Samani, A., & Malekian, A. (2010). The Qanat: a living history in Iran. In G. Graciela Schneier-Madanes & M.-F. Courel (Eds.), *Water and sustainability in Arid regions: bridging the gap between physical and social sciences* (pp. 125–138). Dordrecht: Springer.

Amiri, H. (2001). Resaleh Tariq-e Qesmat-e Ab-e Qalb (in Persian). *Monthly Book of History and Geography, 44–45*, 82–85.

Atkinson, R. (2019). Social cohesion. In A. Orum (Ed.), *Encyclopedia of urban and regional studies* (pp. 1–5). US: Wiley-Blackwell.

Axelrod, R. (1984). *The evolution of cooperation*. New York: Basic Books Inc., Publishers.

Azkia, M., Ghaffary, G., & Hooglund, E. (2011). An analysis of rural studies in Iran. *International Journal of Social Sciences, 1*(1), 17–30.

Balali, R., Keulartz, J., & Korthals, M. (2009). Reflexive water management in arid regions: the case of Iran. *Environmental Values, 18*(1), 91–112.

Baqutayan, S., Ariffin, A. S., Mohsin, M. I. A., & Mahdzir, A. M. (2018). Waqf between the past and present. *Mediterranean Journal of Social Sciences, 9*(4), 149–155. https://doi.org/10.2478/mjss-2018-0124.

Bartold, V. (1971). *Irrigation in Turkistan (Persian translation)*. (K. Keshavarz, Trans.). Tehran: Publication of Social Researches and Studies Institute.

Basati, H., Poursaeid, A., Allahyari, M., Eshraghi, R., & Amin, H. (2019). Social network analysis of local water user associations' actors: evidence from Iran. *Meteorology Hydrology and Water Management, 8*(1), 90–97.

Bazza, M. (2006). *Overview of the history of water resources and irrigation management in the Near East region*. Rome: Food and Agriculture Organization of the United Nations.
Beaumont, P. (1968). Qanats on the Varamin Plain, Iran. *Transactions of the Institute of British Geographers, 45,* 169–179.
Beaumont, P. (1989). In P. Beaumont, M. Bonine, K. McLachlan (Eds.), *The Qanat: a means of water provision from groundwater sources, Qanat, Kariz and Khattara: traditional water systems in the Middle East and North Africa*. London: The Middle East Center, School of Oriental and African Studies, University of London, pp. 13–31.
Boelens, R., Hoogesteger, J., Swyngedouw, E., Vos, J., & Wester, P. (2016). Hydrosocial territories: a political ecology perspective. *Water International, 41*(1), 1–14.
Boucharlat, R. (2003). Iron Age water-draining galleries and the Iranian Qanat. In D. Potts et al. (Eds.), *Proceedings of the first international conference on the archaeology of the U.A.E.*
Bruhn, J. (2009). *The group effect, social cohesion and health outcomes*. Berlin: Springer.
Carballo, D. (2013). Cultural and evolutionary dynamics of cooperation in archaeological perspective. In D. Carballo (Ed.), *Cooperation and collective action: archaeological perspective* (pp. 3–33). Boulder: University Press of Colorado.
De Planhol, X. (2004). *Nomads and pastoralists (Persian translation)* (M. H. Papoli Yazdi, Trans.), Tehran: Samt.
Dede, S. (2011). *From nomadism to sedentary life in Central Anatolia: the case of Risvan Tribe (1830–1932)*. Master thesis, Department of History, Ġhsan Doğramacı Bilkent University, Ankara.
Diamond, J. (2005). *Collapse: how societies choose to fail or succeed*. New York: Penguin Group.
Enayatollah, R., Kouros, G. R., Emam Shooshtari, M. A., & Entezami, A. A. (1971). *Water and irrigation technique in ancient Iran (In Persian)*. Tehran: Water and Power Ministry.
English, P. W. (1998). Qanats and lifeworlds in Iranian Plateau Villages. In J. Albert, M. Bernhadsson, R. Kenna (Eds.), *Transformations of Middle Eastern natural environments: legacies and lessons* (pp. 187–205). Bulletin Number 103. New Haven: Yale University.
Estaji, H., & Raith, K. (2016). The role of Qanat and irrigation networks in the process of city formation and evolution in the Central Plateau of Iran, the Case of Sabzevar. In F. F. Arefian, S. H. I. Moeini (Hg.), *Urban Change in Iran. Stories of Rooted Histories and Ever-Accelerating Developments* (pp. 9–18). Basel: Springer International Publishing.
Fennell, L. A. (2011). Ostrom's law: property rights in the commons. *International Journal of the Commons, 5*(1), 9–27.
Funder, M., Bustamante, R., Cossio, V., Huong, P. T. M., van Koppen, B., Mweemba, C., Nyambe, I., Phuong, L. T. T., & Skielboe, T. (2012). Strategies of the poorest in local water conflict and cooperation—evidence from Vietnam, Bolivia and Zambia. *Water Alternatives, 5* (1), 20–36.
Garoosi, A. (2003). *History of water and irrigation in Province of Kerman (in Persian)*. Tehran: National Commission for Irrigation and Drainage.
Gheybi, M. (2011). Rishe-hāye Farhangi-ye Waqf Dar Miyān-e Zartoshtiān-e Kermān [Cultural roots of Waqf among the Zoroastrians of Kerman]. In A. Sharaf-khani (Ed.), *Waqf: Cheshme-ye Hamishe Jāri [Waqf: the everlasting flowing spring]* (pp. 23–46). Tehran: Sāzmān-e Owqāf Va Omoor-e Kheyriyeh.
Hajian, N., & Hajian, P. (2013). Feasibility of Upgrading Sheikh Baha'i's scroll taking the present situation into account (in Persian). The First National Conference on the Challenges of Water Resources and Agriculture, Isfahan.
Hardin, G. (1968). The tragedy of the commons. *Science, New Series, 162*(3859), 1243–1248.
Hobbes, T. (1962). *Leviathan*. New York: Collier Books.
Janebollahi, M. S. (1990). Water division, transaction and accounting in traditional irrigation in meybod. *Geographical Research Journal, 2*.
Jomehpour, M. (2009). Qanat irrigation systems as important and ingenious agricultural heritage: case study of the qanats of Kashan, Iran. *International Journal of Environmental Studies, 66* (3), 297–315.

Labbaf Khaneiki, R. (1997). Identification of Qanats in Gonabad. In *Archaeological Reports (1)*. Archaeology Research Institute, Tehran.
Labbaf Khaneiki, M. (2014). Genesis of Qanat as a collective response to climate change in Iran (in Persian). *Journal of Asar, Iranian Cultural Heritage Organization, 66.*
Labbaf Khaneiki, M. (2019). *Territorial water cooperation in the central Plateau of Iran*. Switzerland: Springer Nature.
Laessoe, J. (1951). The irrigation system at Ulhu 8th century B.C. *Journal of Cuneiform Studies, 5/1*.
Lambton, A. (1939). The regulation of the waters of the Zayande Rud. *Bulletin of the School of Oriental (and African) Studies, IX*.
Madani, K. (2008). Reasons behind Failure of Qanats in the 20th Century. World Environmental and Water Resources Congress 2008.12.05–16.05.2008, American Society of Civil Engineers, Honolulu, Hawaii.
Madonsela, A. (2017). Sociological perspectives on social cohesion as the principal requirement for social stability. *South African Review of Sociology, 48*(3), 84–98.
Mahmoodian, S., & Qayoomi Bidhendi, M. (2012). Norm of Canals; Water Management System in Safavid Isfahan (in Persian). Architecture and Municipal Engineering Journal, University of Art, No. 10.
Manuel, M., Lightfoot, D., & Fattahi, M. (2018). The sustainability of ancient water control techniques in Iran: an overview. *Water History, 10*(1), 13–30.
Megdiche-Kharrat, F., Ragala, R., & Moussa, M. (2018). Promoting a sustainable traditional technique of aquifers' water acquisition common to arid lands: the case of Ghassem Abad Qanat in Yazd Province (Iran). Water Science and Technology: Water Supply. https://doi.org/10.2166/ws.2018.098.
Moncrieffe, J. M. (2004). *Power relations, inequality and poverty*. London: Overseas Development Institute.
Nash, H., Labbaf Khaneiki, M., & Semsar Yazdi, A. A. (2012). Traditional timing of water shares. In *Proceedings of the International Conference on Traditional Knowledge for Water Resources Management* (pp. 21–23). Yazd, Iran.
Ostrom, E. (1990). *Governing the commons: the evolution of institutions for collective action*. Cambridge University Press.
Papoli Yazdi, M. H., & Labbaf Khaneiki, M. (1998). Water division unit in traditional irrigation systems (in Persian). *Geographic Researches Journal, 49–50*.
Papoli Yazdi, M. H., & Labbaf Khaneiki, M. (2000). Role of qanat in formation of civilizations; theory of sustainability of qanat culture and civilization. In *Proceedings of international conference on qanat, Volume 1*, Yazd, Iran.
Perret, S., Farolfi, S., & Hassan, R. (2006). *Water governance for sustainable development*. London: Earthscan.
Remini, B., Kechad, R., & Achour, B. (2014). The collecting of groundwater by the qanats: a millennium technique decaying. *Larhyss Journal, 20,* 259–277.
Rocha López, R., Boelens, R., Vos, J., & Rap, E. (2019). Hydrosocial territories in dispute: flows of water and power in an Interbasin transfer project in Bolivia. *Water Alternatives, 12*(1), 267–284.
Saedloo, H. (1997). *Water in the civilization of Iran and Islam (in Persian)* (Vol. 17). Rahnamay-e Ketab (Book Guide).
Salimi Moayed, S. (2000). The legend on qanat genesis in Shahdad Region (in Persian). In *Proceedings of International Qanat Conference, Vol. 1*. Yazd: Yazd Regional Water Authority.
Sanaan, B. N. (2020). The qanat system: a reflection on the heritage of the extraction of hidden waters. In C. Hein (Ed.), *Adaptive strategies for water heritage* (pp. 41–57). Cham: Springer.
Semsar Yazdi, A. A., & Labbaf Khaneiki, M. (2017). *Qanat knowledge; construction and maintenance*. Netherlands: Springer.
Semsar Yazdi, A. A., & Labbaf Khaneiki, M. (2019). *Veins of the desert*. Denmark: River Publishers.

Sharifi, A., Pourmand, A., Canuel, E.A., Peterson, L.C., Djamali, M., Lahijani, H., & Naderi, M. (2013). A high-resolution record of climate variability from Neor Lake in NW Iran: Investigating the role of abrupt climate change on human civilization in West Asia. INQUA Quick LakeH workshop on rapidly changing large lakes and human response, Tehran, Iran.

Swyngedouw, E. (2004). *Social power and the urbanization of water: Flows of power.* Oxford University Press.

Swyngedouw, E. (2009). The political economy and political ecology of the hydro-social cycle. *Journal of Contemporary Water Research & Education, 142,* 56–60.

Vahman, F., & Asatrian, G. (2002). *Notes on the language and ethnography of the zoroastrians of Yazd.* Copenhagen: Special-Trykkeriet Viborg.

Walker, B., & Salt, D. (2006). *Resilience thinking: sustaining ecosystems and people in a changing world.* Island Press.

Wilson, S., & Mansour, J. K. (2020). Collective directional movement and the perception of social cohesion. *British Journal of Social Psychology.* https://doi.org/10.1111/bjso.12361.

Yazd Regional Water Authority. (1999). *Basic study on Yazd Province's water resources (in Persian).* Tehran: Iran Ministry of Energy.

Chapter 5
Drought Pump

Abstract Over the past centuries, intermittent droughts in the central plateau of Iran played a crucial role in social and cultural dynamics. The local communities responded to climate change with three different strategies, proportional to the magnitude of water scarcity. During a wet period, they expanded the area of their cultivated lands with a variety of crops and they let less lands lie fallow. When a year turned dry, they shrank the area of farmlands and let more lands lie fallow and resorted to a cropping pattern with lower water demand, and eventually modified their water division systems. With more environmental pressure, they put more focus on local industries, which could supplement their fickle agricultural economy. During the two former stages, population could be still fed and sustained by a mixed economy swaying between agriculture, local industries and trading. However, there was a third stage when aridness crossed the tolerance line of the society, and then even the offset economy did not help. In this case, part of the population began to overflow and migrate to the more favorable regions, spreading their skills already evolved under environmental strains. When a wet period followed the drought, the remaining population gradually returned to agriculture that brought about more food, and as a result population grew again. But the next drought came around and this cycle continued. This cycle can be called "drought pump" whose environmental and cultural mechanism will be examined in this chapter.

5.1 Introduction

Drought-induced migrations have not been limited to history, but they are still occurring nowadays regardless of their cultural dimension. In modern days, the recurring droughts and their societal issues have taken on new dimensions, which may lead to many acute social and political crises and eventually "hydraulic collapse". Here, hydraulic collapse means a situation when socio-economic and political structures fall into an irreversible disorder that is in the first place triggered by water resources crisis.

M. Labbaf Khaneiki, *Cultural Dynamics of Water in Iranian Civilization*,
https://doi.org/10.1007/978-3-030-58900-4_5

Now the growing environmental migration in different parts of the Iranian plateau is alarming. For example, in the province of Southern Khurasan, rural settlements comprise 61.76% of the province's population in 1996, whereas this number dropped to 45.61% in 2011 in the wake of climate change along with other factors (Ebrahim Zadeh and Esmaeel Nejad 2017, p. 16). The eastern and south-eastern provinces were struck by persistent droughts most severely; the provinces of Sistan & Baluchistan, Kerman and Southern Khurasan. According to some studies, since the mid-1990s a dramatic decline in precipitation has affected the Iranian plateau especially the aforementioned provinces, where a decrease of only 1 mm would directly inflict a loss of 1.4 million USD on the local economy (Nobakht et al. 2018, p. 325). In this region between 1996 and 2006, the annual precipitation fell below 100 mm, which may have contributed to a negative net migration rate in over 90% of the towns and cities across the three provinces (Ibid).

In Iran, most of the environmental migrations are attributable to water crisis, but water crisis is not totally down to natural droughts. Nowadays, water mismanagement plays more important role than natural droughts. One of the problems blighting the Iranian water management has much to do with the overlaps and conflicts between different entities, all somehow involved in water. For example, the Iranian Ministry of Energy plans to discourage massive agriculture in the arid regions of Iran, where irrigation efficiency is far below the national standard. It does not make economic sense to cultivate watermelon—for example—in the Iranian desert areas where production of 1 kg of watermelon consumes 300 L of water on average. According to the official reports, more than 2.2 million tons of watermelon is exported from such arid lands, which is tantamount to exporting 50 billion cubic meters of water a year, about 12 times more than the water that all Iranian qanats can discharge during a year. In the face of the real price of water, which never pays off through such agriculture, the Iranian Ministry of Agriculture inclines to develop agricultural activities at any cost (Pouran et al. 2017). Therefore, it is the lack of good water governance which can undermine the sustainable development of Iran.

Water scarcity is bringing the Iranian society to the brink of a socio-environmental catastrophe, including environmental migration. The dimensions of the crisis are so huge that even the Iranian authorities are no longer prudent to be outspoken about it. The Iranian vice president Kalantari recently gave a warning about the looming water crisis, saying that about 50 million people (or about 70% of the Iranian population) are subject to an impending environmental migration as a result of the persistent water scarcity. He said that in 2018 the per capita water supply has dropped to less than 500 cubic meters, whereas an amount less than 1000 cubic meters is considered a critical situation. He has also said that exploitation of renewable water resources should not exceed its 40% according to the global standards, whereas it has exceeded 110% in Iran on average, and in some provinces like Khurasan Razavi it has surpassed 130% (ILNA News Agency 2019).

Although today environmental migration is regarded as an absolute calamity, it was a double-edged sword in pre-modern Iran where most of cultural exchanges were limited to physical movement of people and goods in the absence of such media as TV and internet. Therefore, environmental strains like water scarcity could propel people out of their homelands, opening up more opportunities for them to encounter other cultures. This focus on the positive aspect of environmental migration does not mean that its tragic consequences have been ignored, the consequences that are almost alien to our modern societies. Sometimes, water scarcity and subsequent famines decimated large populations in the Iranian plateau, before they could even seek escape from hunger in other regions. This tragedy can be traced back to the oldest times in the Iranian history. Sassanian king, Peroz (I) who reigned from 459 to 484 AD, had to exempt most of his provinces from annual tax, due to a devastating drought that scorched many farmlands and orchards (Tabari 1972, p. 158). In the years 893 and 894 AD, in the city of Rey, drought was so severe that a 3.8 kg bucket of water was sold at the price of one Dirham (about 12 USD), which was considered very expensive at the time (Ravandi 1985, p. 417).

However, drought was not the only culprit behind such famines. In the past, Iran's agricultural economy was possible only within a very fickle safe zone where all climatic factors like precipitation, temperature and humidity were expected to remain within a short range favorable to farming. For example, if precipitation falls below that limit, people had to go through drought, and if precipitation occurred beyond that limit, flash floods inflicted damage on them. In this case, both scenarios ended up in widespread famines in the Iranian towns and villages. Temperature played the same role. On the one hand, a warm winter increased evaporation and led to less snowfall in the mountains; on the other hand, it could trigger the swarms of such pests as locust that striped the fields. But a cold and long winter could freeze their crops and wreak havoc on their production. Once again, both scenarios could bring about fatal famines. Although, the local communities were equipped with indigenous knowledge to adapt to such changing environmental conditions, environmental pressures sometimes passed the threshold of their resilience.

For example, a freezing cold winter struck the city of Nishapur in 1010 AD when 67 heavy snowfalls hampered all agricultural activities. The surviving crops were all frozen by the next wave of frost and a large population was left bereft of food. That famine killed some 107,000 people in Nishapur (Nafisi 1963, p. 8). The same destiny afflicted Sistan in 1013 AD when a heavy snow blanketed the area and froze many orchards and farmlands (Bahar 1935, p. 360).

Moreover, human-induced disasters contributed to historical famines too. Beyhaghi (1971, p. 609) reports about a severe famine that killed many people in Nishapur in 1040 AD, due to Sultan Mas'ud's mismanagement and incompetence. Āghsarāyi (1983) reports that "in the wake of Oghuz invasion in 1185 AD, the farmlands of Kerman were all destroyed and the city plunged into the abyss of famine. No grain was found at any house, and people resorted to date seeds that they desperately grinded and ate. When they even ran out of date seeds, those hungry people roasted leather shoes and buckets on fire and subsisted on them. Every day some children went missing, where the starving people had no qualms about killing and eating unwatched children. No stray dog and cat remained in the city and they all were caught and eaten by people. It was not rare to see people and dogs wrestling in the streets. If people could beat dogs, they ate dogs, otherwise were eaten by dogs. The emaciated corpses were strewed across the streets, and no

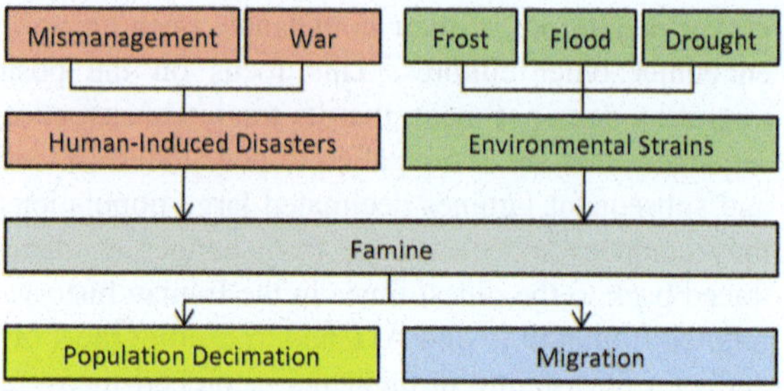

Fig. 5.1 Factors contributing to famine as one of the driving forces behind migration

one was able to bury them". In 1707 AD, Sultan Hosein spent the country's budget on a royal convoy of 60,000 pilgrims who traveled to Mashhad shrine, while the residents of his capital city, Isfahan, were struggling with a catastrophic famine. Some merchants took advantage of Sultan Hosein's mismanagement and fanned the flames by hoarding foodstuffs for more profit (Ravandi 1985, p. 430). In the course of history, human-induced disasters and natural droughts have often gone hand-in-hand in most Iranian communities. The latest example was the "great famine" that was rooted mostly in the political and managerial turmoil at the time of Qajar. According to Majd, about 50% of Iranian population perished between 1917 and 1919 AD when that horrific famine struck the country (2007, p. 85).

Therefore, famine as one of the driving forces behind migration was associated with two groups of factors as environmental strain and human-induced disaster. This relationship is shown in the following figure. Famine could in turn unbalance the ratio of population to available food, and eventually the excess population left the area in the shape of environmental migration. In this chapter, we want to take a closer look at the relationship between drought, famine and migration. Although the other parts of Fig. 5.1 are of great importance, we have to narrow down our argument to better understand how drought is linked to cultural dynamics in the end.

5.2 Literature Review

The deep roots of the Iranian irrigation system are reflected in an array of historical handbooks. The Pahlavi juridical work entitled "Matigan-i Hazar Datistan" devoted a chapter on the qanat system and its legal issues (Bahraman 2014, p. 201–204; Macuch 1981; Shahzadi 2007). The eleventh century historian Abu Sa'id 'Abd a. Hayy Gardizi reports that the Tahirid governor of Khurasan, Abdallah ibn Tahir (d. 844-5), commissioned a book concerning irrigation practices (Kitab-i quni) regulating the legal and practical aspects pertaining to subterranean irrigation channels (Bosworth 2011, p. 45; Labbaf and Semsar 2015; Lambton 1969, p. 217; Rizazada Malik 1995, p. 203). The Arabic treatise by the Iranian author Muḥammad ibn al-Ḥusayn al-Karaji (d. ca. 1029) on the location of groundwater reserves and the construction of underground water channels is extant and describes technical methods that are still valid today (Karaji 1941, 1966). Rashid al-Din Fazlullāh (1247–1318), the powerful vizier of Ghazan Khan, allotted a chapter of his book to different aspects of water issues (1978, p. 11).

The current scholarship on irrigation falls into four groups. The first group is dedicated to the physical evidence and mechanical aspects of qanats, and the implications for traditional water supply, indigenous irrigation and water management systems in Iran (Ahmadi et al. 2010; Amiri 2001; Bartold 1971; Bazza 2006; Beaumont 1968, 1989; Enayatollah et al. 1971; Estaji and Raith 2016; Garoosi 2003; Hajian and Hajian 2013; Janebollahi 1990; Jomehpour 2009; Lambton 1939; Mahmoodian and Qayoomi Bidhendi 2012; Madani 2008; Manuel et al. 2018; Megdiche-Kharrat et al. 2018; Nash et al. 2012; Balali et al. 2009; Saedloo 1997; Honari 1979; English 1998; Semsar Yazdi and Labbaf Khaneiki 2017, 2018, 2019).

The second group of studies concentrates on traditional water management and water division systems in the Iranian highlands (Bonine 1982; Rajabzadeh 2019). Safinejad has explored the aspects of water management and the related social and cultural structures (Safinejad 1989). Jafary and Bradley (2018) focus on the challenges farmers face in the groundwater irrigation management in Central Iran, while Molle et al. (2004) investigate water rights in the village of Jalalabad in Isfahan province. Also, a monograph has been devoted to the principal concepts of traditional water management, different types of water division systems and hydro-social organizations across Iran (Labbaf Khaneiki 2007).

The third group of studies sheds light on the impact of environmental factors on water management and water resources to account for the great variation in qanat systems across the Iranian highlands. These studies highlight the different morphologies of qanats, the different techniques used in their construction and maintenance, as well as the relationship between various geographies and specific water management systems (Labbaf Khaneiki and Semsar Yazdi 2015; Papoli Yazdi and Labbaf Khaneiki 2004; Papoli Yazdi et al. 2000; Semsar Yazdi et al. 2005). Labbaf Khaneiki (2014) understands the genesis of qanat as a collective response to climate change in Iran. Christensen touches upon hydrological changes that affected water management systems and then contributed to Tahirids' decision to change the capital of Khurasan from Marv to Nishapur (2016, p. 236). Although this group of studies bears more resemblance to this research, they are still devoid of an interpretation of how drought could have affected cultural dynamics.

5.3 Climatic Turbulence and Ancient Human Settlements in Iran

After the Wurm glaciation, some 12,800 years ago, the earth happened to return to a semi ice age condition. This freezing change is known as Younger Dryas that persisted for 1200 years (Sobooti 2011, p. 112). At the time, the central plateau of Iran enjoyed some permanent surface streams originating from the surrounding elevations and running down toward some sporadic lakes and water bodies, paving the way for prehistoric human settlements.

Over the prehistoric period, the Central Iran especially the regions today on the edge of the Iranian deserts were overrun with lakes, springs and rivers according to geological and archeological evidences. The banks of these rivers and lakes accommodated many sedentary farming communities and triggered the creation of some civilizations there. Thousands of years elapsed and small patches of desert grew and intruded into the prosperous areas, and accordingly the water bodies and rivers came to dwindle and then dried up, subjecting the settlements dependent on those waters to a great change. But the settlements which nestled on the banks of the big rivers far away from the intrusion of the central desert continued their growth into metropolises and principle civilizations. The famous civilizations of Mesopotamia like Babylonian and Akkadian, which relied on the river of Euphrates give credence to this claim. Also the river of Karkhe that has survived the desertification could quench the lands of Shoosh city (Susa) for a long time, and made it possible for this ancient city to evolve into the hub of Ilam civilization and then Achaemenian headquarters (Sykes 1984, p. 48–49).

However, the destiny of the settlements located on the margin of the Iranian desert went down a different path. In ancient times, the people of Sialk irrigated their farmlands with the plentiful of water from the natural spring of Fin (Namani 1979, p. 434) and those who lived in Teppe Hesar in Gonabad, Teppe Siyah and Teppe Bayas Abad much probably relied on the river of Kal Shoor. The settlement of Teppe Hesar, once populated between the third and first millenniums BC, was situated on the southern bank of Kal Shoor. Kal Shoor is a gully crusted with salt, which once watered Teppe Hesar and now sometimes brings a seasonal flood from the surrounding elevated lands. The both banks of this gully are so saturated with salt that even the most salt-resistant desert plants cannot live there. What really became of those civilizations?

This story has happened to many of the ancient settlements on the edge of the central desert whose suitable water sources vanished in the wake of climate change. The drying up of prehistoric rivers and lakes led to the annihilation of some human settlements like what Gabriel (1992) came across in the Lout Desert, but some others took a different strategy to survive. They invented and developed the qanat technique to better adapt to such a harsh condition, wherever the environmental circumstances allowed.

Even nowadays, in the central plateau of Iran many of the population centers are situated on the coasts of the gone prehistoric lakes according to geological evidences. One of those lakes was once covered all the area around the swamp of Gav Khooni in Isfahan. This lake has left three levels of sedimentation whose highest level is situated at the height of 1550 m above the sea level. In other words, the water level was some 80 m higher than what we see in the swamp of Gav Khooni in its full capacity. One can imagine the city of Isfahan located right on its coast, and the other towns like Ezheh, Harand and Varzaneh are submerged. In the province of Yazd, the lowland of Abarkooh was not the exception, though its meteorological conditions were different from that of the northern and western areas. At least two levels of sedimentation are perceptible along the road from Abarkooh to Taft, and the town of Abarkooh has been built just on the highest level of sedimentation (Ramesht 2001, p. 104). According to Gabriel there once existed a relatively big lake in the central lowland of Iran (Gabriel 1992, p. 311). On the western side of this lowland close to the assumed lake such cities as Nayin, Aghda, Ardakan, Meybod and Yazd nestle. The city of Nayin is located at the highest level, Aghda at the second level and Ardakan at the third level and actually at the bottom of the assumed lake. There were also other lowlands—once ancient lakes—parallel to the aforementioned lake, whose coasts favored the formation of the primitive human settlements. In fact, most of the lakes in the central plateau of Iran were not deep enough to form any particular geological stratification, but the type of sedimentation across those lowlands implies early existence of shallow and relatively stagnant lakes whose coasts were the cradle of ancient settlements (Ramesht 2001, p. 105). But the climate change gradually made those water bodies dry up, leading to a completely different environmental landscape.

Some paleo-climatological studies show that the dry, humid, warm and cold climates intermittently prevailed over the Iranian highlands, resulting from a variety of factors from changes in sun energy to changes in wind patterns (Sharifi et al. 2013). One of the dry periods took place in the late third millennium BC especially in the southern parts of the Iranian highlands according to a new archaeological study on Sisatn region. That dry period posed a great challenge to the region's water resources and sparked a massive environmental migration from Sistan to the mountainous areas as far north as Alborz Mountains (Mortazavi 2016).

The archeological findings prove that the climate change at ancient times broke out at a relatively high pace. As an instance, the vestiges left from Scythians who dwelt north of Iranian plateau between the second and first millennium BC suggest that their territory was then covered with lush forests (Maleki 1998, p. 176). Also in Birjand east of Iran, the stone carving of Lakh Mazar exhibits the images of some animals like lion that have long become extinct in the area. This carving dates back to the first millennium BC, when the region enjoyed a better climate suitable for that fauna (Labbaf Khaneiki and Bashash 1994, p. 29). But this area was devoured by the advancing desertification, and the inhabitants resorted to exploiting groundwater to make it through the new harsh condition. The climate change could not wipe the human presence off the central Iran, but it altered the basis of life and production and gave rise to "qanat civilization" which was built upon the groundwater extracted by qanat system.

An interesting study done on the Lake Van in eastern Turkey can shed light on the climatic condition of ancient times and its impact on the human civilizations. The sediments that have settled at the bed of this lake over the past 14 years can provide a clue to estimate the climatic conditions at the ancient times. The ratio of oxygen isotope measured in the two layers of aragonite and calcite revealed the climatic conditions before and after the Roman Empire. Therefore the highest relative humidity in the Middle East has been dated between 8400 and 4100 years ago. Afterwards a dry period began some 4100 years ago and culminated in a long-standing drought about 2100 years ago. Ever since, many smaller fluctuations occurred in the amount of humidity (McCormick 2012, pp. 169–220). Between 7000 and 6000 years ago, the Iranian plateau was fortunate to sit in between two rich systems; Indian Ocean monsoon which affected its southeast, and North Atlantic sourced rains which reached the northwest of the region. However, later monsoon rains began moving southward and fading out, and North Atlantic system receded and gradually left the region drier (Jones et al. 2013, p. 31).

Such fluctuations in the amount of precipitation left its imprint in the history of ancient civilizations in the Middle East. For example, the Akkadian empire collapsed around 2170 BC, which can be associated with their changing climate and its impact on their socio-economic structures (Diamond 2005, p. 183). The same climate change recurred in seventh century AD, which affected a large area from Mesopotamia to the Mediterranean. An abrupt increase in temperatures, the shrinking of water resources, the invasion of the sand dunes and eventually the collapse of agricultural production systems were all brought about mostly by that climate change, which set the stage for the Arabs' successful incursion (Issar 1995).

Arash Sharifi and his colleagues have conducted a similar research on Neor Lake northwest of Iran, which has come up with almost the same result as that of the Lake Van study. They have dated different layers of sediment at the bottom of Neor Lake by gauging the ratio of oxygen 18 to oxygen 16. The oxygen with the mass number of 18 is heavier than the oxygen 16 that is much more abundant in nature. A molecule of water which consists of oxygen 18 is accordingly heavier than a molecule with oxygen 16 that is also easier to evaporate. The sea creatures with calcium carbonate shell like crabs, shrimps, scallops, and so on absorb those molecules of water containing oxygen 18 that remains in their shell. Those animals die and are buried under the sediments where the paleo-climatologists can date them by measuring the radioactive decay of the heavier isotope of oxygen. During dry periods when evaporation rate is higher, the water with oxygen 16 more easily evaporates and enters into the atmosphere, and as a result more atoms of oxygen 18 remain in the lake and then gets concentrated in the sediment layer (Rohling 2013, p. 916). Sharifi's measurement implies that around 4000 and then 2000–2500 years ago, the Iranian plateau went through drier years (Sharifi et al. 2013).

If we compare such paleo-climatological information to the historical records, we can come up with a more realistic image of the situation of water resources in the past. For example, we know that amid such climatic fluctuations, the twelfth and thirteenth centuries AD stand out as a warm and dry period when the Iranian highlands experienced a harsher condition (Sharifi et al. 2013). The Mongolian invasion that swept away the agrarian society of Iran in 1219 AD is partially attributable to climatic changes and the ensuing droughts at the time (Gautam 2009, p. 39). Also, many historical records bear witness to the above-mentioned paleo-climatological fact. As an instance, in 1275 AD, a Persian geographer named Qazvini has described the town of Abarkooh (Yazd province) as a town where it never rained (1994, p. 192). Also another geographer named Ḥamavi in the same thirteenth century has mentioned in his book that the town of Abarkooh was bare and treeless and no garden was visible on its outskirts (2004, p. 82). Moreover, Ibn al-Balkhi in twelfth century has mentioned the qanat system as the only water source for Abarkooh (1995, p. 291). Therefore, we can come to the conclusion that some 800 years ago the region enjoyed a warmer and drier climate, which acknowledges the paleo-climatological facts.

However, some 200 years after Ibn Balkhi, in the fourteenth century another geographer named Hafiz Abroo reported that Abarkooh enjoyed a river in addition to its qanats as saying that "the town enjoys a moderate weather and its water is supplied by both river and qanat, and its condition is favorable for fruits and grains" (Hafiz Abroo 1996, p. 111). What Abroo reports does not much resemble the town that Hamavi describes as dry and treeless. It seems that in the twelfth century, the aridity was in full swing, but the climate later tended to become wetter, and paleo-climatological evidences give credence to this assumption as well.

5.4 Proto-Industrialization and Virtual Water

Such environmental limitations as climatic turbulence and their impacts on the development of industrial activities have always been subject to academic debates. The theory of "proto-industrialization" first appeared in Franklin Mendels's Ph.D. dissertation and was then introduced to the literature of development in 1972 (Mendels 1972). According to this theory, prior to modern industrialization in the west, in Europe between sixteenth and nineteenth centuries kind of spontaneous organic industrialization came into existence as a result of interactions between geographical conditions and socio-economic predispositions in those regions (Catheren 2008, p. 1). Those pre-modern industries came to expand their influence area and worked their way into the markets even out of their local territories. Proto-industrialization was rooted in the rural regions of Europe, and is regarded as the first stage of the modern industrial revolution in the west (Mendels 1972, p. 241). Proto-industrialization is premised on the fact that the climatic and geographical conditions in many European regions did not favor full involvement in agricultural activities all year round, so the manpower temporarily detached from agriculture could set the stage for rural industries to flourish. Capital was gradually built up in rural industries, which could lead to more workforces being drained from agricultural sector, and such a cycle brought about more development for rural industries.

Development of rural industries in turn challenged many of norms and mechanisms in the European societies, and the traditional systems started to loosen their hold, the systems that used to be very conservative and systematically keep the population growth in balance with the available economic resources. As a result, proto-industrialization pieced together all the essential elements needed for the launch of the modern industrialization such as workforce, capital, liberalism, technical expertise, commercial and manufacturing enterprises, inter-regional markets, and so on, and thus the economic destiny of the west headed down a new path.

Nevertheless, the theory of proto-industrialization was modified by some scholars like Levine who regarded proto-industrialization as part of a bigger process referred to as "proletarianization". Levine believes that proto-industrialization was followed by an abrupt population growth, which disturbed the traditional patterns of land ownership and social structures in rural communities. That process left behind a considerable population who no longer possessed enough farmlands for their livelihood and had to work for others in return for wage. Therefore, workforce was transformed into a proletarian nature in the wake of proto-industrialization, which eventually gave rise to the modern industrial revolution (Levine 1977). Later some scholars like Kreidte tried to conflate the theories of Mendels and Levine in order to reach a description on historical move of Europe from feudalism to capitalism (Kreidte et al. 1981, p. 8). In all these interpretations, workforce or in fact human capital remains as a key factor for industrialization process (Suarez-Villa 1985, p. 319).

The same pattern of proto-industrialization has historically been in place in the central plateau of Iran, though it could not trigger a modern industrial revolution due to a set of socio-political impediments. Proto-industrialization is anchored in an interaction between geographical, social and economic processes that may unfold even far away from the European lands. For example, proto-industrialization also came about in Japan and later made it possible for modern industrialization to thrive there (Howell 1992).

Even in the central plateau of Iran, proto-industrialization gave rise to a capacity for modern industrialization through an interaction between geographical, social and economic factors, though this capacity was overshadowed by the cultural hegemony of the surrounding areas and was stifled from its beginning. In fact, qanat as the only water supply for agricultural production was bound up with its own technical limitations. This hydraulic system is impaired by some natural, technical and managerial impediments, which are all associated with the structural peculiarities of qanat as shown in Fig. 5.2.

Qanat can drain out only the overflow of groundwater based on gravity, the same water that comes down in the shape of rain or snow at the top of some sparse elevations in the region. Therefore, the area of farmlands was not able to keep pace with any increase in population, due to such a limitation in water supply (qanat). Qanat's water cannot be increased in proportion to the increasing water demand. The qanat tunnel is not typically higher than 160 and wider than 80 cm, into which groundwater seeps slowly, builds up and then flows down. Therefore, it is

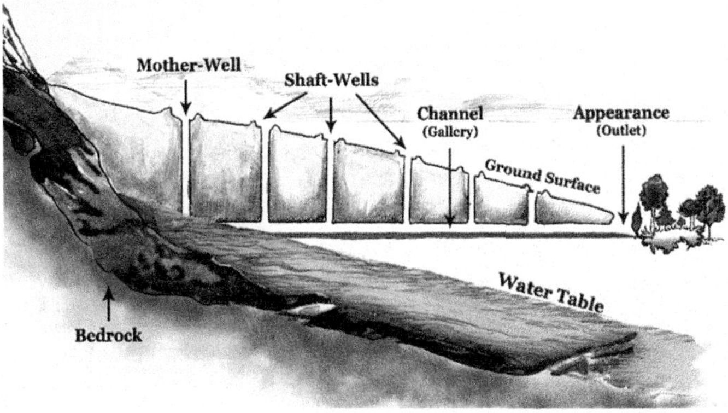

Fig. 5.2 Longitudinal section of a typical qanat (Semsar Yazdi and Labbaf Khaneiki 2019)

impossible to obtain a huge water flow through a typical qanat. Although its water flow can be increased to some extent by digging further into the aquifer or building side branches (Semsar Yazdi and Labbaf Khaneiki 2017), other factors may keep such measures at bay.

For example, geological condition is not usually homogenous in a particular region, so the extension of a tunnel or its side branches may come across impassable barriers such as saline formations or hard rocks, which rule out further advance. Moreover, the extent of aquifer and its hydraulic gradient are among the natural limitations, which hinder more advances into the aquifer. Besides, the legal bound of neighboring qanats should be observed while extending the qanat gallery through water bearing zone. In many cases, there are other qanats adjacent to a particular qanat, all taking water from the same aquifer. Therefore, a qanat advancing toward another qanat can cause a decrease in its water flow, and so is forbidden by tradition and law. Even extending of a qanat should be carried out in tandem with its neighboring qanats, which is calculated by a qanat expert based on the depth of each qanat.

The cost of qanat extension can be another limitation. Qanat extension can increase the groundwater infiltration area along the tunnel and accordingly results in more water flowing out of qanat. Most of the qanats enjoy petty land ownership, owned by low-income farmers who do not afford the high expense of such constant extension. Therefore, qanat extension is a measure taken to retain the qanat flow or retrieve the same flow, if qanat flow would dwindle. In other words, qanat extension serves to retain the present situation, and is not aimed at obtaining more water than the normal hydraulic regime of qanat.

It should be also noted that qanat is usually used for surface (flood) irrigation with traditional earthen ditches in which a considerable portion of water percolates or evaporates. Therefore, the more they expand their cultivated lands, the more they have to build earthen open ditches, the more water would escape through percolation and evaporation, and the less is the irrigation efficiency.

Moreover, ownership structure of qanat impedes any more development in its agricultural system. Qanats mostly enjoy a common ownership, and heredity makes water shares become smaller and smaller over time. On the other hand, ownership of water independent of land is considered another limitation for the development of qanat-based agricultural systems.

These factors altogether prevent qanat-based agricultural systems from expanding for the purpose of feeding the growing population, the population that increases exponentially, but the available water resource remains constant at best. Such limitations that are inherent in the qanat-based agricultural systems do not allow the whole manpower to remain involved in agricultural sector, and pave the way for other economic sectors to play their role to supplement the income deficiency of such a subsistence agriculture. Small local industries have been among the most important economic sectors in the central plateau of Iran, which have always served as the cornerstone of the financial and commercial processes in the course of history.

Qanat-based agricultural systems used to end up in the agricultural products that were all usually consumed in this same region. But a considerable portion of the industrial products were intended for exporting to as far other regions as China and India. One can liken those oases in the Iranian desert to some islands scattered across an ocean. Each island has a fragile agricultural system based on irrigational technique of qanat, which cannot provide the booming population with enough food. Hence, the islanders have to hone their skills in industrial and trading activities in order to barter their products for more food with other regions. Those islands have no next neighbors, but they have to travel a relatively long distance across the ocean to reach the first residential land. For example, at the time of Safavids (1501–1736 AD), Yazd became the hub of silk textile that could rival the Chinese products. Silk textiles were exported from Yazd to as far east as India, China, Mongolia and Turkistan (Tashakori Bafghi 2013, p. 54).

This example implies some similarities between the conditions that prevailed in the Iranian oases and in some islands like England. Therefore, in the central plateau of Iran proto-industrialization led to the expansion of economic and commercial activities across the geography. This current of goods and capitals resulted from proto-industrialization in turn gave rise to some social consequences.

In Europe, long winters put limitation on constant agriculture and drove people to invent a way to supplement their income using the surplus manpower from agricultural sector. In the central plateau of Iran, the same process took place but because of water limitation rather than long freezing winters. Thus, in the central plateau of Iran small home industries could take advantage of this condition and thrive to a considerable extent. These industries played a crucial role in the economic structure of the region, the industries that rose to national and sometimes international fame such as Persian carpet, textile, silk, pottery, tile, glass, metal products, etc.

The subsistence agriculture based on qanat irrigation could not satisfy all needs of the growing population, and could not absorb the entire available workforce either. Therefore, the workforce surplus resorted to local industries whose products were usually exported to other regions from where many agricultural products were imported. In fact, they imported water in the shape of agricultural products, taking into account the fact that they no longer needed to worry about water by importing the crops that were already watered somewhere else. A particular amount of water was required to produce a particular amount of crop, which is called virtual water. For example, production of 1 kg of grains required between 1000 and 2000 L of water, whereas production of 1 kg of cheese needed as much as between 5000 and 5500 L of water (Hoekstra 2002, p. 13). Hence, they could actually bring 5000–5500 L of water in the shape of virtual water by importing only 1 kg of cheese.

Proto-industrialization not only resulted in importing virtual water as an adaptation strategy against environmental strains, but also diversified the local economy through developing a variety of small industries. The people who were involved in those small industries enjoyed more mobility compared to those who were tethered to their farmlands. Hence, those craftsmen were more susceptible to migration, when environmental pressures passed their resilience line. They migrated to other regions and took with them the skills that proto-industrialization had bestowed on them.

5.5 Conclusion

The story of the human's evolution may inspire us to find some subtle connections between the intermittent droughts, waves of migration and cultural interaction with other territories. It is said that millions of years ago, the Sahara region in Africa went through some recurring dry periods with wet intervals when the precipitation rose, some fresh water bodies came about and a green savanna covered the landscape. During the wet period, the favorable conditions of the region drew many species in droves including humans. However, these appealing conditions did not last long and the species were driven away toward the four geographical directions by a severe dry period that turned the savanna into a harsh impassible desert. The desert served as a barrier that kept the members of a particular species separated from each other for a while long enough for them to evolve in different ways in order to adapt themselves to the new geographies to which they had migrated. In fact, the diversity of life is simply indebted to a cycle of wet Sahara, migration into Sahara, dry Sahara and then migration out of Sahara. Of course, several waves of primitive humans' migration out of Africa are associated with this cycle, which is called "Sahara Pump" (Parravicini and Pievani 2016).

It was very probable that over the past centuries, recurring droughts in the central plateau of Iran played a similar role but in a cultural rather than biological manner. In other words, the dry periods pushed the population surplus out of the qanat-dominated regions when the environmental pressure even went beyond the threshold that the offset economic sectors no longer sufficed to recruit the excess workforce. The local communities responded to the climate change with three different strategies, proportionate to the magnitude of water scarcity. During a wet period, they expanded the area of their cultivated lands with a wider variety of crops and they made less lands lie fallow. In Yazd, in addition to private farms, there were some communal farms locally named "Lard" which were collectively cultivated only during wet years when water was available more than needed by the private lands. When a year turned dry, they shrank the area of farmlands, let more lands lie fallow and resorted to a cropping pattern with lower water demand. Also they put more focus on handicrafts and local industries which could make up for the income that they lost to such a downturn in agriculture. As mentioned, this stage was akin to proto-industrialization that could supplement their fickle agricultural economy.

In the central plateau of Iran, proto-industrialization used to unfold in a simple manner. Most of the raw materials were imported from other regions, processed and then exported to other regions with added value. For example, cocoons were imported to Yazd from the Caspian coasts whose environmental condition lent itself to breeding silkworms (Tashakori Bafghi 2013, p. 54). Raw cocoons were turned into silk threads and textiles and then were exported to Bombay in India (Labbaf Khaneiki and Semsar Yazdi 2015, p. 163). Hence, the people of Yazd could produce silk without using their precious water to irrigate vast mulberry orchards. Their silk products were sold in India and the earned money was used to buy foodstuff from Fars province. This way they could convert "skill" to "food" without toiling away in the farmlands.

The same economic route as that of silk was established for other products such as henna, sesame, asafetida, rock candy and even opium (Azizi and Karimian 2015, p. 245). Sesame seeds as raw material were obtained from other agricultural regions, and were transformed to different products like sesame oil and tahini. The sesame products were exported to other regions and their revenue was spent on foodstuff from other regions again. Even today some 80% of national sesame products come from Yazd, while 70% of sesame seeds are being imported from other countries like Afghanistan, Pakistan and Somalia (Malek Mohammadi 2018, p. 3). Sesame needs to be irrigated 20 times over its cropping period (Fallah et al. 2015, p. 13), a high water demand that is difficult to be afforded by the farmers of Yazd. Hence, they have always preferred to bring its seeds from other regions and profit from its added value. The same logic applied to henna production. At the time of Naser al-Din Shah (1831–1896 AD), some 3.5 million kilograms of henna was exported from Yazd to many regions across Iran every year (Hoseini 2004, p. 159), though a single shrub of henna was not grown in Yazd (Ibid., p. 158). Most of the henna leaves grown in the regions Bam, Shahdad, Jiroft, Manoojan (Kerman province), Minab, Bandar-Abbas (Hormozgan province), and Roodbar (Gilan province) were transported to Yazd to produce that reddish-brown dye (Ibid., p. 159).

During the two former stages, the population could be still fed and sustained by a mixed economy swaying between agriculture, local industries and trading. However, there was a third stage when the aridness crossed the tolerance line of the society, and then even the offset economy did not help. In this case, part of the population began to overflow and migrate to the more favorable regions as far as Khurasan, Fars, Khuzestan, etc. There have been several important waves of migration from for example Yazd to other regions in the wake of some intermittent droughts. That is why it is very easy to find the communities of Yazd's people in almost every city around the central plateau of Iran, and accordingly it is quite possible to track down the impact of Yazd's art and culture on the host societies.

According to a census taken in 1956, majority of migrants living in the city of Mashhad were from Yazd. They built a new street named Yazdi-Abad most of whose inhabitants were originally from Yazd. Those migrants were mostly merchants or craftsmen who were involved in tailoring, bakery, weaving, bath keeping and qanat building (Najaf-zadeh 2010, p. 132). Najaf-zadeh (2013) shows how migrants from Yazd contributed to economic prosperity of Mashhad ever since the period of Safavids through cultural exchanges. Some groups of migrants who left Yazd in the 1880s, headed toward northeast of Iran that enjoyed better environmental conditions (Yusofifar and Shaa'bani 2017, p. 30). According to Sunderland (1968, p. 615), water scarcity has always served as a driving force behind social mobility of this kind in Yazd region, which was sometimes orchestrated even by the landlords who moved some of their tenant farmers from one property to another of theirs or even from agriculture to other economic activities at the time of drought. Hence, the unstable water supplies have given rise to many dynamic settlements whose population was moved from one place to another over time in search of water. This dynamics has formed a specific landscape on the Iranian plateau, which can easily mislead Europeans mistaking many of those abandoned villages in a given area for a general decline in the population and the economy (Christensen 2016, p. 68). When a water source—usually a qanat—ran dry, the inhabitants did not vanish, but they changed their economic strategy to adapt to the new condition by shifting from horticulture to local industries and trading or by migrating to other places.

Meybod is a small town near Yazd, whose potteries are well known all over the country. The Iranians coveted those beautiful potteries with their amazing white paintings in a blue glazed background. There is a village in the province of Khurasan Razavi, whose handicraft is mostly a type of pottery that bears a striking resemblance to that of Meybod. The name of the village is Mend, probably a shortened pronunciation of Meybod. In the village, some elderly allege that their ancestors came from Meybod in Yazd long time ago. Other than those similar potteries and people's anecdotes, no reliable evidence is available to prove that those villagers are really the holdover of a wave of migration provoked by a severe drought. Nonetheless, we can still point out many communities in other regions, who were driven out of their qanat-dependent regions and brought with them the fruits of proto-industrialization to the host societies (Labbaf Khaneiki 2019).

Ghiyas al-Din was an adept textile designer from Yazd, who could gain fame at the Safavid court. At the time of Shah Abbas I (1571–1629 AD), Ghiyas al-Din moved to Isfahan and had a great influence on textile industry there (Tashakori Bafghi 2013, p. 61). Like his fellow people, he was far from happy with the harsh conditions of Yazd, which eventually drove him out of his hometown along with his skills, as he complained in a poem of him as follows:

"Whoever lives in Yazd is desperate.

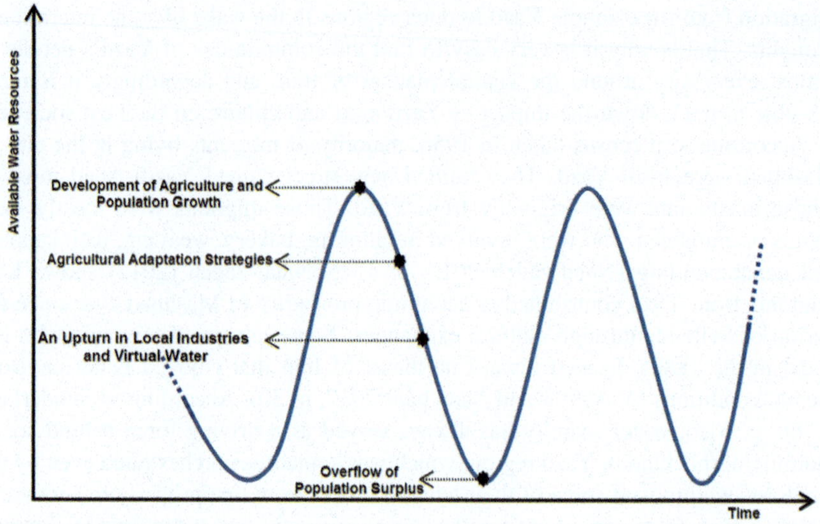

Fig. 5.3 Drought pump formed by an interaction between changing water resources and adaptation strategies

Whoever is textile designer is more desperate.
Those interested in poetry are worse off than both of them.
How desperate I am who live in Yazd as a textile designer and poem[1]" (Maserrat 2018, p. 18).

Figure 5.3 schematically shows how water scarcity could have pumped the population surplus out of the region, giving rise to a socio-environmental process that has always contributed to the cultural evolution of Iran.

References

Āghsarāyi, M. (1983). Tārikh-e Salājagheh [History of Seljuqs]. (O. Toorān, Ed.). Tehran: Asātir.

Ahmadi, H., Nazari Samani, A., & Malekian, A. (2010). The Qanat: A living history in Iran. In G. Schneier-Madanes, M. F. Courel (Eds.), *Water and sustainability in arid regions: Bridging the gap between physical and social sciences* (pp. 125–138). Dordrecht: Springer.

Amiri, H. (2001). Resāleh-ye Tariq-e Qesmat-e Ab-e Qalb [A treatise on the method of water division]. *Ketāb-e Māh-e Tārikh Va Joghrāfiyā, 44–45,* 82–85.

Azizi, H., & Karimian, H. (2015). Naghsh-e Bāzargānān Dar Bāft-e Tārikhi-ye Shahr-e Yazd Dar Asr-e Qājār [Role of merchants in the historic fabric of Yazd City in the period of Qajar]. *Tārikh-e Eslām, 16*(1), 231–261.

[1]Bichāre kasi ke shahr-e Yazdash vatan ast
Bichāre-tar ānk-e naghsh-bandish fan ast
Zin har do batar har ānke ahl-e sokhan ast
Nāchār kasi ke har se dārad cho man ast

Bahar, M. T. (1935). *Tārikh-e Sistān [History of Sistan]*. Tehran: Zavvār.
Bahraman, F. (2014). Mātigān-i Hazār Dātistān [Book of one thousand laws] (S. Oryan, Trans.). Tehran: Elmi Publication.
Balali, R. M., Keulartz, J., & Korthals, M. (2009). Reflexive water management in arid regions: The case of Iran. *Environmental Values, 18*(1), 91–112.
Bartold, V. (1971). Ābyāri Dar Turkistān [Irrigation in Turkistan] (K. Keshavarz, Trans.). Tehran: Enteshārāt-e Moasese-ye Tahghighāt Va Motālea't-e Ejtemāyi.
Bazza, M. (2006). *Overview of the history of water resources and irrigation management in the near east region*. Rome: Food and Agriculture Organization of the United Nations.
Beaumont, P. (1968). Qanats on the Varamin Plain, Iran. *Transactions of the Institute of British Geographers, 45*, 169–179.
Beaumont, P. (1989). The Qanat: A means of water provision from groundwater sources. In P. Beaumont, M. Bonine, & K. McLachlan (Eds.), *Qanat, Kariz and Khattara: Traditional water systems in the middle east and North Africa* (pp. 13–31). London: The Middle East Center, School of Oriental and African Studies, University of London.
Beyhaghi, A. (1971). Tārikh-e Beyhaghi [Beyhaghi's history] (A. A. Fayyāz, Ed.). Mashhad: Dāneshgāh-e Ferdowsi Mashhad.
Bonine, M. (1982). From Qanāt to Kort: Traditional irrigation terminology and practices in central Iran. *Iran, 20*(1982), 145–159.
Bosworth, C. E. (2011). *The ornament of histories*. London: BIPS Persian Studies Series.
Catheren, O. S. (2008). *Protoindustrialization*. England: University of Cambridge.
Christensen, P. (2016). The Decline of Iranshahr; irrigation and environment in the middle east 500 BC–AD 1500 (S. Sampson, Trans.). I.B.Tauris & Co Ltd.
Diamond, J. (2005). *Collapse: How societies choose to fail or succeed*. New York: Penguin Group.
Ebrahim Zadeh, I., & Esmaeel Nejad, M. (2017). Climate refugees the future challenge for regional transitions; case study southern Khurasan (in Persian). *Journal of Development and Geography, 48*(3), 1–18.
Enayatollah, R., Kouros, G. R., Emam Shooshtari, M. A., & Entezami, A. A. (1971). *Āb Va Fann-e Ābyari Dar Irān-e Bāstān [Water and irrigation technique in ancient Iran]*. Tehran: Vezārate Āb Va Bargh.
English, P. W. (1998). Qanats and lifeworlds in Iranian Plateau Villages. In J. Albert, M. Bernhadsson, & R. Kenna (Eds.), *Transformations of middle eastern natural environments: Legacies and lessons* (pp. 187–205, Vol. 103). New Haven: Yale University.
Estaji, H., & Raith, K. (2016). The role of Qanat and irrigation networks in the process of city formation and evolution in the central plateau of Iran, the Case of Sabzevar. In F. F. Arefian, & S. H. I. Moeini (Hg.), *Urban change in Iran. Stories of rooted histories and ever-accelerating developments* (pp. 9–18). Basel: Springer International Publishing.
Gabriel, A. (1992). Oboor Az Sahāri-ye Irān [Passing through persian desert] (F. Najd Samiyi, Trans.). Mashhad: Āstān Qods-e Razavi.
Garoosi, A. (2003). *Tārikh-e Āb Va Ābyāri Dar Ostān-e Kermān [History of water and irrigation in province of Kerman]*. Tehran: Komite-ye Melli-ye Ābyāri Va Zehkeshi.
Gautam, P. K. (2009). Climate change and the military. *Journal of Defense Studies, 3*(4), 37–48.
Ḥafiẓ Abroo, (1996). *Ǧuġrāfīyā-i Ḥāfiẓ Abrū [Ḥāfiẓ Abrū's geography]* (Vol. 2). Tehran: Mīrāth-i Maktūb.
Hajian, N., & Hajian, P. (2013). *Emkān-e Berooz Kardan-e Toomār-e Mosoom Be Sheykh-e Bahāyi Bā Tavajjoh Be Sharāyet-e Mojood [Feasibility of upgrading Sheikh Baha'i's scroll taking the present situation into account]*. Isfahan: Anjoman-e Melli-ye Ābyāri Va Zehkeshi-ye Irān.
Ḥamavi Baghdādī, Y. (2004). Muʿjam al-Buldān [Dictionary of countries] (A. Monzavi & A. Mehr-Parvar, Trans.). Tehran: Sāzmān-i Mīrāṯ-i Farhangī-i Kishvar, 2004.
Hoekstra, A. Y. (2002). Virtual water: An introduction. In *Proceedings of the International Expert Meeting on Virtual Water Trade*. IHE Delft, The Netherlands.

Honari, M. (1979). *Qanats and human ecosystems in Iran* (PhD Dissertation). Retrieved from Dissertation and Theses Database, University of Edinburgh.

Hoseini, M. (2004). Vaziyat-e Eghtesādi-ye Yazd Dar Dowre-ye Nāser al-Din Shāh [Economic situation of Yazd at the time of Naser al-Din Shah]. Dovvomin Hamāyesh-e Melli-ye Irān-shenāsi (pp. 137–175). Tehran: Bonyād-e Irān-shenāsi.

Howell, D. L. (1992). Proto-industrial origins of Japanese capitalism. *Journal of Asian Studies, 51* (2), 269–286.

Ibn al-Balkhi (1995). Fārsnāma. Shiraz: Persian Study Foundation.

ILNA News Agency. (2019). Kalantari's warning on the country's water situation (in Persian). Donyaye Eghtesad Newspaper, January 2019, No. 4519.

Issar, A. S. (1995). Climatic change and the history of the middle east. *American Scientist, 83*(4), 350–355.

Jafary, F., & Bradley, C. (2018). Groundwater irrigation management and the existing challenges from the farmers' perspective in central Iran. *Land, 7*(15), 1–22.

Janebollahi, M. S. (1990). Nizām-e Taghsim Va Hesābrasi Va Kharid Va Foroosh-e Āb Dar Ābyāri-ye Sonnati-ye Meybod [Water division, transaction and accounting in traditional irrigation in Meybod]. *Faslnāme-ye Tahghighāt-e Joghrāfiyāyi, 2*(1990), 52–100.

Jomehpour, M. (2009). Qanat irrigation systems as important and ingenious agricultural heritage: Case study of the Qanats of Kashan, Iran. *International Journal of Environmental Studies, 66* (3), 297–315.

Jones, M., Djamali, M., Stevens, L., Heyvaert, V., Askari, H., Noorollahi, D., & Weeks, L. (2013). Mid-Holocene environmental and climatic change in Iran. In A. Petrie Cameron (Ed.), *Ancient Iran and its neighbours: Local developments and long-range interactions in the 4th millennium BC.* Oxbow Books.

Karaji, M. A-Ḥ. (1360/1941). Kitab al-inbat al-miyah khafiya [Extraction of the hidden waters]. Hyderabad: Matba'at al-Dairat al-'Uthmaniyah.

Karaji, M. A-Ḥusain (1966). Istikhrāj-i Ābha-i Pinhāni [Extraction of the hidden waters] (H. Khadivjam, Trans.). Tehran: Intisharat-i Bunyad-i Farhang-i Iran.

Kriedte, P., Medick, H., & Schlumbohm, J. (1981). *Industrialization before industrialization: Rural industry in the genesis of capitalism.* New York: Cambridge University Press.

Labbaf Khaneiki, R., & Bashash, R. (1994). *Lākh Mazār-e Birjand.* Tehran: Iranian Cultural Heritage Organization.

Labbaf Khaneiki, M. (2007). *Nezāmhā-ye Taghsim-e Āb Dar Irān [Water division systems in Iran].* Tehran: Ganjine-ye Melli-ye Āb-e Irān.

Labbaf Khaneiki, M. (2014). Peydāyesh-e Qanāt Pāsokhi Be Taghirāt-e Eghlim Dar Falāt-e Irān [Genesis of Qanat as a collective response to climate change in Iran]. *Journal of Asar, Iranian Cultural Heritage Organization, 66*(2014), 77–84.

Labbaf Khaneiki, M., & Semsar Yazdi, A. A. (2015). Gardeshgari-ye Qanāt [Qanat tourism]. Yazd: International Center on Qanats and Historic Hydraulic Structures (UNESCO-ICQHS).

Lambton, A. (1939). The regulation of the waters of the Zayande Rud. *Bulletin of the School of Oriental (and African) Studies, 9*(3), 663–673.

Lambton, A. (1969). *Landlord and peasant in Persia.* Oxford: Oxford University Press.

Levine, D. (1977). *Family formation in an age of nascent capitalism.* London: Academic Press.

Macuch, M. (Ed.). (1981). *Das sasanidische Rechtsbuch "Mātakdān i hazār dātistān".* Wiesbaden: Steiner.

Madani, K. (2008). *Reasons behind failure of Qanats in the 20th century.* World Environmental and Water Resources Congress 2008. 2008, May 12–16, American Society of Civil Engineers, Honolulu, Hawaii.

Mahmoodian, S., & Qayoomi Bidhendi, M. (2012). Hanjār-e Mādi: Nezām-e Modiriyat-e Āb Dar Esfahān-e Safavi [Norm of canals; water management system in Safavid Isfahan]. *Nāme-ye Memāri Va Shahr-Sāzi, 10*(2012), 141–153.

Malek Mohammadi, F. (2018). Yazd Sokkān-dār-e Tolid-e Panjāh Noa' Farāvarde-ye Konjedi-ye Keshvar [Yazd at the Helm of the production of fifty sesame products in the country]. *Samt, 2277*(959), 3.

Maleki, H. (1998). *Forests and desertification in Iran*. Tehran: Ayandeh Publication.
Manuel, M., Lightfoot, D., & Fattahi, M. (2018). The sustainability of ancient water control techniques in Iran: An overview. *Water History, 10*(1), 13–30.
Maserrat, H. (2018). Khājeh Ghiyās-e Naghsh-band, Honarmandi Tavānā Va Shāeri Dānā [Ghiyas al-Din the designer, an adept artist and a wise poet]. In M. Kangarani (Ed.), *Majmooa'e Maghālāt-e Naghsh Va Naghshband [Anthology of articles on textile design and designer]* (pp. 11–28). Tehran: Farhangestān-e Honar.
McCormick, M., et al. (2012). Climate change during and after the roman empire: Reconstructing the past from scientific and historical evidence. *Journal of Interdisciplinary History, xliii*(2).
Megdiche-Kharrat, F., Ragala, R., & Moussa, M. (2018). Promoting a sustainable traditional technique of aquifers' water acquisition common to arid lands: The case of Ghassem Abad Qanat in Yazd Province (Iran). *Water Science and Technology: Water Supply, 19*(2), 527–535.
Mendels, F. (1972). Proto-industrialization: The first phase of the industrialization process. *Journal of Economic History, 32*(1972), 241–261.
Molle, F., Mamanpouth, A., & Miranzedh, M. (2004). *Robbing Yadullah's water to irrigate Saeid's Garden. Hydrology and water rights in a village of Central Iran*. In IMWMI Research Report. Band 80. Colombo: International Water Management Institute.
Mortazavi, M. (2016). *Gozāresh-e Bāstan-shenāsi-ye Sevomin Fasl-e Haffāri Dar Tappeh Dasht-e Sistān [Archaeological report on the third season of excavation in Teppe Dasht, Sistan Region]*. Zahedan: Cultural Heritage Organization of Sistan and Baluchistan.
Nafisi, S. (1963). *Dar Pirāmoon-e Tārikh-e Beyhaghi [About Beyhaghi's history]* (Vol. 1). Tehran: Forooghi.
Najaf-zadeh, A. (2010). Pajooheshi Darbāre-ye Aghvām, Aghaliyat-hā Va Mohājerin-e Sāken-e Shahr-e Mashhad [Ethnicities, minorities and immigrants based in Mashhad]. *Mashhad-Pajoohi, 3*(4), 115–143.
Najaf-zadeh, A. (2013). *Gorooh-hāye Mohājer Be Mashhad Az Safaviyeh Tā Konoon [Groups of migrants coming to Mashhad from Safavid Period to the present]*. Mashhad: Shorā-ye Eslāmi-ye Shahr-e Mashhad.
Namani, F. (1979). *Takāmol-e Feodālism Dar Irān [Evolution of feudalism in Iran]* (Vol. 1). Tehran: Khārazmi.
Nash, H., Labbaf Khaneiki, M., & Semsar Yazdi, A. A. (2012). Traditional timing of water shares. In *Proceedings of the International Conference on Traditional Knowledge for Water Resources Management*, Yazd, Iran, pp 21–23.
Nobakht, R., Ghasemi, A., & Gholami, M. (2018). Analysis of demographic-economic consequences of drought in the eastern regions of Iran (in Persian). *Journal of Geography and Regional Planning, 8*(1), 313–327.
Papoli Yazdi, M. H., & Labbaf Khaneiki, M. (2004). *Qanāt-hāye Taft [Qanats of Taft]*. Mashhad: Entershārāt-e Papoli.
Papoli Yazdi, M. H., Labbaf Khaneiki, M., Labbaf, K. R., & Jalali, A. (2000). *Qanāt-e Ghasabeh Yek Ostooreh [Qanat of Qasabeh as a Myth]*. Mashhad: Sherkat-e Āb-e Mantaghe-yi Khorāsān Razavi.
Parravicini, A., & Pievani, T. (2016). Multi-level human evolution: Ecological patterns in Hominin phylogeny. *Journal of Anthropological Sciences, 94*(2016), 167–182.
Pouran, R., Raghfar, H., Ghasemi, A., & Bazazan, F. (2017). Calculation of economic value of virtual water from the standpoint of irrigation efficiency (in Persian). *Iran Applied Economic Researches Journal, 6*(21), 189–212.
Qazvini, Z. b. M. b. M. (1994). *Relics of lands and news of people (in Persian)* (1st edn). Tehran: Amirkabir Press.
Rajabzadeh, H. (2019). Hashtād-o Hasht Sanad-e Ab-e Keshāvarzi Dowre-ye Qājār [Eighty-eight documents on irrigation from Qajar Iran]. Tokyo: University of Tokyo (Institute for Advanced Studies on Asia).
Ramesht, M. H. (2001). Daryāche-hāye Dowrān-e Chāhārom Bastar-e Tabalvor Va Gostaresh-e Madaniyat Dar Irān [Quaternary lakes: Ground of formation and development of civilization in Iran]. *Faslnāme-ye Tahghighāt-e Joghrāfiyāyi, 16*(60), 90–111.

Rashid al-Din Fazlullāh (1978). Latāyef al-Haghāyegh. (G. Taher, Ed.) (Vol. 2). Tehran: University of Tehran.
Ravandi, M. (1985). *Tārikh-e Ejtemāyi-ye Irān [Social history of Iran]* (Vol. 5). Tehran: Katibeh.
Rizazada Malik, R. (Ed.). (1384/2005). Zain al-akhbar. Tehran: Anjuman-i Athar va Mafakhir-i Farhangi.
Rohling, E. J. (2013). Oxygen isotope composition of seawater. In S. A. Elias (Ed.), *The encyclopedia of quaternary science* (Vol. 2, pp. 915–922). Amsterdam: Elsevier.
Saedloo, H. (1997). Āb Dar Tamadon-e Irān Va Eslām [Water in the civilization of Iran and Islam]. Rahnamāy-e Ketāb (Vol. 17).
Safinejad, J. (1989). *Nezām-hāye Ābyāri-ye Sonnati Dar Irān [Traditional irrigation systems in Iran]* (Vol. 2). Mashhad: Āstān Qods Razavi.
Semsar Yazdi, A. A., & Labbaf Khaneiki, M. (2017). *Qanat knowledge; construction and maintenance*. Netherlands: Springer.
Semsar Yazdi, A. A., & Labbaf Khaneiki, M. (2019). *Veins of the desert*. Denmark: River Publishers.
Semsar Yazdi, A. A., & Labbaf Khaneiki, M. (2018). Qanāts of Iran: Sustainable water supply system. In F. Sulas & I. Pikirayi (Eds.), *Water and society from ancient times to the present* (pp. 219–237). England: Routledge.
Semsar Yazdi, A. A., Labbaf Khaneiki, M., & Dehghan, B. (2005). *A survey on the Qanats of Bam from technical and engineering point of view*. Tehran: UNESCO Tehran Cluster Office.
Shahzadi, R. (1386/2007). Qanun-i Madani-i Zartusthiyan [Civil law of Zoroastrians]. Tehran: Faravahar.
Sharifi, A., Pourmand, A., Canuel, E. A., Peterson, L. C., Djamali, M., Lahijani, H., & Naderi, M. (2013). A high-resolution record of climate variability from Neor Lake in NW Iran: Investigating the role of abrupt climate change on human civilization in West Asia. In *INQUA Quick Lake Workshop on Rapidly Changing Large Lakes and Human Response*, Tehran, Iran.
Sobooti, Y. (2011). *The warm earth the result of 21st century*. Tehran: Geography and Cartography Institute of Gitashenasi.
Suarez-Villa, L. (1985). Industrialization in the development world: Process cycles and the new global division of labour. *Canadian Journal of Regional Science, VIII*(3), 307–331.
Sunderland, E. (1968). Pastoralism, nomadism and the social anthropology of Iran. In W. B. Fisher (Ed.), *The Cambridge history of Iran, Volume I; The land of Iran* (pp. 611–683). UK: Cambridge University Press.
Sykes, P. (1984). History of Persia (S. M. T. Fakhr Dayi Gilani, Trans.) (Vol. 1). Tehran: Donyaye Ketab.
Tabari, M. (1972). Tārikh al-Rosol Va al-Molook [History of the prophets and kings] (S. Nasha'at, Trans.). Tehran: Bongāh-e Tarjomeh Va Nashr-e Ketāb.
Tashakori Bafghi, A. A. (2013). Darāmadi Bar Sana'at-e Bāfandegi-ye Yazd Dar Asr-e Safavi [An introduction to textile industry in Yazd in the period of Safavids]. *Pajoohesh-Nāme-ye Tārikh-e Ejtemāyi Va Eghtesādi, 2*(1), 51–65.
Yusofifar, S., & Shaa'bani, R. (2017). Barresi-ye Vaziyat-e Mohājerat-e Dākheli Dar Dowre-ye Qājār Bar Asās-e Gozāresh-e Rooznāme-ye Vaghāyeh al-Ettefāghiyeh [Situation of internal migration in the period of Qajar according to the reports of Vaghāyeh al-Ettefāghiyeh Newspaper]. *Tahghighāt-e Tārikh-e Ejtemāyi, 7*(2), 23–47.

Chapter 6
The Water Delusion

Abstract This chapter examines some traditional rituals and customs, which revolve around water and its footprint even in the deepest corner of the Iranians' collective unconscious. The chapter tries to come up with some interpretations on why and how water has manifested itself in the rituals of Iran's local communities.

This chapter recounts some rain-claiming ceremonies that are a significant part of what we call intangible water culture. Those ceremonies are expressive of the cosmology of local people on the one hand, and of their relationship with water on the other hand. All those rain-claiming ceremonies fall into four categories: defiance, supplication, appeasement and combined actions. The notion behind all ceremonies is an omnipotent phenomenon like a supernatural master that mysteriously wields power over rain. Local people may take different strategies to align the rain master with their own needs. Sometimes locals consider the rain as a right that has been withheld from them at the time of drought. Hence, they do something weird in defiance of social, moral or even religious norms in order to put pressure on the rain master so he relents and gives the rain back to them. Sometimes they entreat the rain master to have mercy on them and compassionately quench their thirst, or they may appease him by throwing him a sop. In some other ceremonies, defiance, supplication and appeasement may be practiced in combination.

This chapter examines "qanat (Qanat is a gently sloping subterranean canal, which taps a water-bearing zone at a higher elevation than cultivated lands.) marriage" and its socio-economic function as well. In some regions of Central Iran the locals used to hold a wedding ceremony for their qanat (water gallery) when it came to dry up. They asked a widow in the village to become the wife of qanat and dwell in a house built just close to the qanat exit. The villagers treated the qanat's wife as a real bride, and all the customs were observed the same as that of a real wedding. This chapter also explains why a qanat was considered a male being capable of marrying humans, and why the locals took such a custom seriously.

M. Labbaf Khaneiki, *Cultural Dynamics of Water in Iranian Civilization*,
https://doi.org/10.1007/978-3-030-58900-4_6

6.1 Introduction

For our ancestors nothing was as mysterious as water. They could not restrain their thirst for water or find any substitute for it, and water could take their lives if they drowned. They could not cultivate and grow their food without water, and they were again doomed to starvation if the flood waters washed their crops away. Water came down from the infinite sky in the shape of rain and snow, and no cause could be imagined behind the rains but dozens of deities reclining on the pillows of cloud. Our ancestors became intrigued by the measureless seas and oceans whose horizon mysteriously petered out in the sky, as if they emanated from the endless divinities. The permanence of the big rivers stimulated their imagination to attribute such streams to different gods or goddesses who generously quenched their thirst, and they held the same divinities responsible for the floods that were believed to be a manifestation of god's rage and revenge as a punishment for their sins. In the arid lands, water served as a line of communication through which gods and people could interact. A drought or a flood was considered a bad omen that god was not content with the people's deeds, and a bounteous river or qanat signified that god was rewarding them for their complete obedience. Gods and people talked in the language of water.

Therefore, Zoroaster, the ancient Iranian prophet, was convinced that the entire creation in fact originated from water (Abbasi et al. 2017, p. 118). According to his cosmology, Ahura-Mazda, the supreme god, entrusted the important task of creating water to Khordād, a high-ranking divinity of "perfection" (Boyce 1985, p. 27). Khordād later took care of his creation with the help of Anahita who was known as the divine guardian of water (Nusservanji Dhalla 1938, p. 176). Zoroaster's dualistic cosmology was manifested in an everlasting battle between Apaosha, demon of drought, and Tishtrya, divinity of rainfall (Eyvazi 1966, p. 153). The entire universe was seen as a battle field where Ahriman, the destructive spirit, and Ahura-Mazda endlessly fought by pitting their creatures against each other, of which the most obvious fight was between Apaosha, created by Ahriman, and Tishtrya, created by Ahura-Mazda. The true believers should not have stood idle, but they were expected to take sides by praying for Tishtrya's victory. Hence, water and drought were not viewed as simple natural phenomena, but they reflected a power relation between the evil and the good spirits on a cosmological scale. Water was a mirror for them to exhibit what was going on behind the scenes in their world.

Water provided them with answers to satisfy their curiosity about creation and the very first moment of being. As mentioned in Chapter one, Zoroastrian Akhshij philosophy (Ghaemi et al. 2009, p. 64) describes water as one of the four fundamental elements of creation along with earth, air and fire. They believed in a direct correspondence between human body and heavenly world as a macrocosm (Stausberg and Vevaina 2015, p. 245). Ancient Persians used to sacrifice to the water as an instrument for divine purification (Ibid., p. 132). The sanctity of water was rooted in Zoroaster's own teachings according to which the creation of water was assigned to the guardianship of Khordād (Boyce 1985, p. 27). The same sanctity has remained in the Iranians' mentality, despite many cultural upheavals and religious conversions that took place in the course of Iranian history (Javadi 2013). Even nowadays, in many Iranian villages, water is still respected and for example, the practice of spitting or urinating directly in a water stream is utterly abhorrent (Rasooli 2015).

Water has established a special relationship between Iranians and their environment, a relationship that underlies an intricate nexus of social, cultural and economic structures. This nexus is manifest not only in their irrigation techniques, but also in many intangible cultural elements such as customs and beliefs.

6.2 Rain-Claiming Ceremonies

One of the important examples of intangible water culture is the ceremonies performed to make rain. Ancient Iranians used a complicated symbolism to invoke the soul of rain against drought. For example, in the archeological site of Tepe Hissar[1] (south of Damghan, Semnan province), the ancient potteries were adorned with the figure of a goat enjoying two big curved horns, which is believed to have been a symbol of rain (Ghazanfari and Karimi 2016, p. 62). The same symbol was painted on the potteries of Susa in the fourth millennium BC, depicting a goat whose long curved horns have embraced a circular farm (Aliabadi 2012, p. 10). Rain ceremonies and rites vary from region to region, though some of them are strikingly similar across a vast geography. This similarity has tempted some scholars to contend that such rites share a mythological root dating back thousands of years. According to ancient Persians, the entire world was an arena where Tishtrya (God of Rain) and Apaosha (Demon of Drought) fought an everlasting battle. Tishtrya's defeat could usher in a hard time of drought, famine and hunger (Moshiri 2014, p. 136). Hence, true Zoroastrians were obliged to side with Tishtrya against Apaosha by any means, which were manifested in a variety of prayers, customs and rites. Some of the ceremonies still practiced across Iran may be a holdover from those ancient times. Regardless of their mythological origins, all the rain-claiming ceremonies fall into four categories: defiance, supplication, appeasement and combined actions, as follows (Fig. 6.1).

6.2.1 Defiance

Sometimes locals consider the rain as a right that has been withheld from them at the time of drought. Hence, they do something weird in defiance of social, moral or

[1]Tepe Hissar was populated between the fifth and the second millennium BC.

Fig. 6.1 The fourth millennium BC pottery of Susa , showing a goat whose horns have engulfed a crop, as the symbol of rain supporting agriculture (Masoomi 1970)

even religious norms in order to put pressure on metaphysical powers to give the rain back to them. This kind of mentality in relation to the divinities has not been limited to a particular place or a particular time, as al-Jāhiz (776–869 AD) reports about the Arabs before Islam for example. "They performed a ceremony called nār al-estemtār in which a cow was tortured to compel the deities to bring rain, because cow was considered a blessed animal. They took a cow to a hilltop and set fire to the firewood tied to its tail and horns. The animal ran wild in despair, and those people looked up as the sky in the hope of rain" (al-Jāhiz 1920, p. 466).

This kind of rain-making ceremony is commonplace across the Iranian plateau. For example, in the village of Kalkhoran in Ardabil, when it does not rain enough, a group of young women (sometimes men) parade along the streets, each with a wooden spoon in one hand and a small flint stone in the other, hitting them together to make noise. They chant together:

> "What does Ms. Chamcheh want?
> She wants rain from God.
> She knocks at every door.
> She begs for a little charity".

If their trick does not work and rain does not show up, this time a group of elderly women take off their headscarves and put a wooden yoke on their shoulders while walking around in the village, in the face of the Islamic rules that necessitate wearing headscarves for women and respecting the elderly for all. If it does not rain yet, the villagers come together and lever a very big boulder into a river named Ali Dervish as the last resort. If it starts raining, they manage to move the boulder back to its original place (Fig. 6.2).

Taking an innocent hostage is another act of defiance against God, which may make him relent and release the rain according to the locals. For example, in the

Fig. 6.2 A group of people marching along the streets, wielding spoons and making strange noises to make rain (by Kiana Labbaf Khaniki)

village of Dehbala, Meymand, Fars Province, all the men and boys come together and walk along the streets while hitting two pebbles together and chanting a special song. They knock at the doors on their way, where someone comes out, pour some water on them, and then give them some dry fruits. They put all the dry fruits in a sack and dump them all in a water spring as wastefully as possible to anger the metaphysical powers that are deemed responsible for their calamity. Afterwards, three persons among the crowd suddenly start running away, and the others chase them, until they would be caught and dragged back to the same water. The crowd pushes the three persons into the water and attempt to drown them, until someone steps forward and mediates by promising that it would rain if the three would be spared. The crowd let go of the three persons in the hope that God sends down rain for the sake of that promise and those innocent people (Fig. 6.3).

In Fariman, Khurasan Razavi province, seyyed men bear the brunt of the lack of rainfall. Seyyed people are those who are believed to be descended from Prophet Mohammad. They choose one of them, who is the most revered by the public, and forcibly sit him down on the floor in the village mosque in the direction of kiblah. People drench him with buckets of water or even with garden hose, pushing him to pray for water in such an awful condition (Papoli Yazdi and Jalali 1999, p. 195). In Paweh, Kermanshah province, this practice takes a different shape. They swing a holy man by his limbs and throw him into a pool of water in order to shame God into bringing rain (Tavakoli 1975, p. 69) (Fig. 6.4).

Sometimes this custom turns more violent. For example, in Mamasani tribe, a group of people comes together and sings a local song that asks God to send down rain. They go up to everyone to scrounge for foodstuff that would be sold in the end or bartered for dates. They hide a small round stone inside the flesh of one of those dates and then give out all the dates among the people. They watch to see who

Fig. 6.3 Pouring water on people to make rain (by Kiana Labbaf Khaniki)

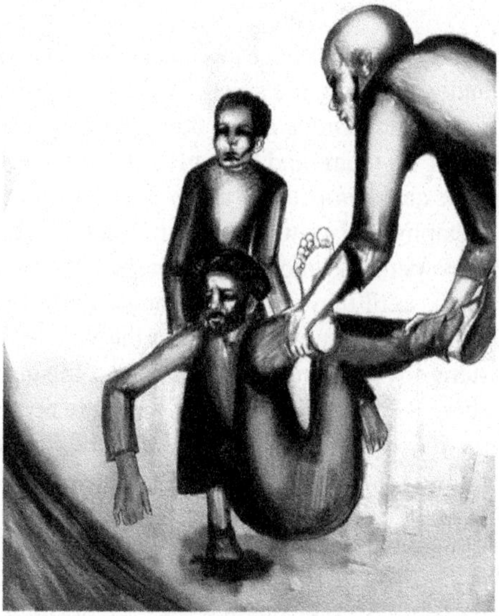

Fig. 6.4 Throwing an innocent man into water to elicit rain (by Kiana Labbaf Khaniki)

would take the date that contains the stone. The mob goes for the unlucky person and starts beating him up, until someone steps in and acts as surety for him, promising on God's behalf that it would rain if they let go of their victim. They stop beating on condition that it would rain in three days, otherwise they find the surety himself and beat him up unless another man stands surety for him and this cycle continues until God gives in and sends the rain down (Mirnia 1989, p. 224). This custom is widespread and can be found across the Iranian highlands as diverse as Sarvestan in Fars province (Homayooni 1992, p. 412), Joopar in Kerman province

(Khazayi 2015, p. 159), Zarand in Kerman province (Ibid., p. 179), Birjand in south Khurasan province (Ibid., p. 172) and Yasuj in Kohgiluyeh and Boyer-Ahmad province (Mirnia 1989, p. 223).

The same ceremony is practiced in Rabor, Kerman province with some deviations. At the time of drought, someone would be singled out to wear a shepherd costume, paint his face black and hang a small bell from his neck. He is called "Louk". He walks around in the streets, having such an awkward appearance, and a group of villagers follows him, while chanting:

"God! Give us rain.

Give endless rain!

Give wheat to the landlords.

Give barley to the donkey-owners.

Give millet to the hen-owners.

Hey Louk, hey Louk".

As soon as Louk reaches someone's house, the house-owner pours water on his face and gives a handful of wheat flour to his followers. The villagers gather all the flour and drop a green bead in it, while making dough. They knead the dough and then bake loaves of bread to be given out among the villagers. They carefully watch to find out in whose bread the green bead would turn up. They put the blame for drought on whoever finds the bead inside his bread. They drag him to the field, tie him up to a tree and whip him hard as long as someone else steps in and stands surety for him. If it does not rain, the surety would be tied up to the tree and whipped. If the whipping does not work and the rainy clouds do not show up after a few days, the first person who had found the green bead in his bread would be whipped again (Khazayi 2015, p. 129).

Another version of this ceremony is common among Baseri tribe, Fars province. To make rain, one of the best young men is chosen for the ceremony. He wears a felt coat awkwardly, putting his feet instead of his hands into the sleeves. The coat is fastened tightly around his waist with a leather belt, a bell is dangled from his neck, a turban was put on his head and he is sarcastically called "old man". The setting of this ceremony is intentionally envisioned against the social norms. He pretends to be sad while walking around and stopping by every tent, along with a group of people who sing a traditional song as follows:

"We took the winds away.

We brought rain.

We had a handful of barley that we sowed on the top of a hill.

God! Give us rain for the herders.

Give us rain for the family men".

At this point, the young man suddenly leaps into the air with joy, and shouts out:

"I just brought rain. I want a reward in return".

Each tent gives him something like a handful of rice, flour, sugar, tea, date or some oil. The tent owner asks him that "hey old man! When would it rain?" He replies that "Just in seven days".

Those who accompany the young man gather all the flour and then bake traditional breads called "Soolok". They bury seven small stones in the breads and then distribute the breads among the group. They wait to see who is first to find the stone in his mouth. They grab him and beat him up, until someone intervenes and stands surety for him, promising that it would rain in seven days. When it finally rains, a group of children sings together the following song:

"It's raining, rain of spring.
My brother is riding a hound!
He's riding a rabid horse!
He's going to fight a jackal!
The jackal says that he's drunk.
A broken sword in the jackal's hand!
If I would have loved God, I should have killed the blind judge"!

The above-mentioned song sounds like an encrypted statement, as if it does not make sense at all. Maybe some readers tend to interpret such songs to discover deeper layers of their culture, rooted in their collective unconscious. But I prefer to take this song at face value, and conclude that those people use even such surreal words as an act of defiance against the logic of language to spite the divinity who is expected to surrender to their awkward behaviors, in order to obtain more rain. Language is another social norm that has been put in place by the same divine being that now should loosen its grip on the rain, otherwise that norm would no longer be observed.

In the village of Seyf Abad, Sarvestan, Fars province, a thin-bearded man called "Kooseh" is picked for this ceremony. Many kids gather around Kooseh who wears an artificial white beard, hangs a bell from his neck, and wears a felt coat, with two big horns on his head and a stick in his hand. Kooseh walks around being followed by the kids. Kooseh stops at each door, where the kids sing together:

"It will rain, if God wills.
Give us rain, God!
Bless us, God!
For a loaf of bread!"

Afterward, Kooseh rings the bell, the house-owner opens the door, and everyone enters, while singing:

"Kooseh is Nader Shah[2]!

---

[2]Nader Shah Afshar was one of the most powerful kings in Iranian history and of course bearded, ruling from 1736 to 1747.

It's raining on the rooftops.

I had a farm.

I sowed a land.

We silly kids hoped for God's blessing."

"Please help me", Kooseh rings the bell several times and then says in a sad voice. "What's wrong with you old man", asks the house-owner. Kooseh answers: "a flood came and took my children. All my jewels went down a fish's throat". The house-owner gives him some money, but pours water on Kooseh and his companions when they leave the house. Kooseh goes to the market and buys some tea and sugar, and then makes a campfire in the field, on which he pretends to dry his clothing, though it is not wet at all, sarcastically saying that: "what a torrential rain! I got drenched"! On the same fire, Kooseh makes tea for the kids. But he asks the kids to drink their tea hot and without sugar. Kooseh beats up the first one who complains about the tea, until another kid stands surety for him and assures Kooseh that it would rain in a few days. Using sarcasm is the main part of this ceremony, when Kooseh complains about a flood and dry his clothing on fire on a dry rainless day.

Similar ceremonies are practiced in the village of Samiyeh, Borazjan, province of Bushehr, but with some differences. The men get together and pick someone as a bride who is called Galin. However, their bride is male, who wears a felt cloak and a long white beard made of sheep wool. The male bride ties a rope around his waist and wears a straw hat to which two goat horns are attached. He hangs a bell from his neck and whitens his face with wheat flour. They put a bridle on the bride and drag him from street to street by his bridle, while singing:

"Our bride is so ugly.

He's ploughed many lands.

Raindrops are big.

His horn is golden.

Come on! Hurry up Golden horn bride!"

The men who follow the male bride beg for foodstuff from every house on their way. Sometimes the house-owner ambushes them from the rooftop and surprisingly pours a bucket of water on them as soon as they knock at the door. In the end, all men sit together and the male bride divides the foodstuff among them, but he stealthily hides a small stone inside one of those shares. One of the men jumps up and runs away as soon as he finds the stone in his food. Other men chase him to catch and beat him up. But he runs to someone's house and takes refuge there. The house-owner stands surety for him and promises that it would rain in a few days, if they release him. The men accept the surety and disperse. After a few days, if it would not rain, the men gather around the same house again and angrily claim for rain. If the house-owner repeats his promise and acts as surety again, the men give more time to the unlucky man, otherwise they look for him and beat him up. The man may take refuge in another house and it continues until it rains (Seddiq 1975,

p. 56). This ceremony has some elements that apparently challenge the social norms, like choosing a man as bride and even exaggerating his masculinity by putting a wool beard and a felt cloak on him. It is also very strange to put a bridle on a human, whereas only donkeys and horses are bridled. Putting the blame on innocent people and beating them up for a drought can be interpreted as defiance against the divinities who are out of their reach but still responsible for both social norms and their calamity.

In many regions in the Iranian central plateau, the act of defiance revolves around a donkey's skull. Many scholars have tried to trace this custom back to the pre-Islamic cultures. For example, some argue that ancient Persians used to sacrifice donkeys for God Mehr, since the killing of cows was considered abhorrent according to Zoroaster (Bahar 1998, p. 301). Another interpretation that is more convincing attributes this custom to an ancient mythology about a three-legged donkey. According to Zoroastrian mythology, the three-legged donkey is a giant creature that always lives in the middle of Farākh-kard Sea. When this donkey urinates in the sea, all the waters would be purified, otherwise the infections and diseases would take over the world. The three-legged donkey is believed to assist Tishtrya—a Zoroastrian benevolent divinity—in collecting and dividing the water of seas (Carnoy 2004, p. 18). It is possible that a ceremony whose focal point is donkey would be a remainder of an ancient belief that venerated a legendry three-legged donkey for its positive role in water distribution (Papoli Yazdi and Jalali 1999, p. 203). However, what the villagers do with the donkey's skull is more of a disrespectful mockery than treating a helpful sacred divinity. Therefore, I think that there may be something simpler behind a donkey ceremony, though it bears some resemblance to Tishtrya's friend, the three-legged donkey. The villagers simply want to annoy the deities and distract them from gripping the rain, by disturbing the social and cultural norms in which the skull of a dead donkey is absolutely useless let alone embellishing it and showing it off.

In the village of Nazl Abad, Razavi Khurasan province, the women find a donkey's head and put makeup on it as if they do it for a real bride. When they finish with the makeup, they stick the head on the tip of a cane and carry it out of the village to a water well. One of the women plays a frame drum (daf) while following the donkey's head whose garish makeup is so hilarious. The crowd shout the word "Shabash" that means "cheer up", which seems irrelevant to their predicament because of the rainless season. When they reach the well, they drop the donkey's head down the well in the belief that it will rain. In this performance, different elements are intentionally put in a contradictory order maybe to challenge the social orders. They go for the head of donkey that is less valued than the other farm animals, and they offer a dead one that is even more useless. They put human makeup on the donkey's face, which makes it sillier and more laughable and can be very offensive to the deities if it plays the role of a sacrifice. Playing a happy drum and cheering up can be sarcastic at the time of a calamity when all crops are wilting since it does not rain.

In Mashhad, Razavi Khurasan province, sometimes the villagers go beyond the head of a dead donkey, but they put garish make-up on a living donkey and then they take it to a public bath the same as they do with a real bride (Khazayi 2015, p. 192). This ceremony can be an attempt to show the divinities to what extent their social norms have been messed up by the lack of rain.

A weirder ceremony is practiced in the village of Bostaq in Ferdows, South Khurasan province, which can really spite the divinity believed to be responsible for drought. The village women find a donkey skull and then put a garish make-up on it the same as they do with a bribe. Afterwards they put the skull in a coffin that is carried on their shoulders to the cemetery. They put down the coffin on an old grave where an immature girl stands right by the coffin. The women pour some water down inside the girl's shirt, so that the water would spill from under her skirt, as if she urinates in the coffin and on the skull. If it means a simulated urination on a grave (a taboo), it would be so offensive to the divinity that the rain would be unleashed for fear of further offences that can plunge the world into a real chaos. Then they put the skull on fire and dance together around it to a frame drum, in a region where women's dancing in public is religiously reprehensible. Doing so, they believe that it would rain.

Women's dancing is used to spite the patriarchal divinity, especially where the women have religious sacred names. In the village of Kalateh Soleyman in Birjand, South Khurasan province, villagers stick a donkey's skull on the tip of a cane and carry it around in the village. They paint the skull will garish colors, and put some grass between its jaws. Some women play frame drums (daf) and seven women all named Fatimah publicly dance to the music (Khazayi 2015, p. 165). We should bear in mind that in South Khurasan province, women's dancing in public is religiously frowned upon, let alone the women named after Fatimah who is deeply venerated by Shiite Muslims (Fig. 6.5).

Sometimes the villagers go even further and do something worse to spite the religious figures. For example, in the village of Nimvar, Mazandaran province, there is a sacred lake that is believed to have been blessed and spiritually owned by Khidr. Khidr has been mentioned in the Quran as a righteous man of God, who enjoys and teaches a mystic knowledge. At the time of drought when the villagers want more rain, they dump garbage in the lake to outrage Khidr who expects everyone to respect his lake. They believe that Khidr would shower them with torrential rain shortly thereafter to take revenge. When they have enough of rain, they go back to the same lake and clean it up to stop the rain (Ibid., p. 394).

In the village of Jilu, Nishabur, Razavi Khurasan province, rain-claiming ceremony ends up in vandalism. As an attempt to annoy the divinity who has withheld rain from the village, they sneak into the homes of all old women whose ovens lie in the direction of kiblah (mecca), and then they stealthily destroy their ovens. To do so, some people knock at the old lady's door and keep talking to her to divert her attention, while some others discreetly go inside and destroy her oven (Khazayi 2015, p. 269). They may believe that destroying all ovens belonging to the old ladies especially those which are in the direction of sacred kiblah can make heaven surrender and send rain.

Sometimes the act of defiance takes weirder shape. In the village of Eshaq Abad, Nishabur, people believe that it would rain, if they write down on a piece of paper

Fig. 6.5 Dancing of seven women all named Fatimah in a conservative Muslim community as an act of rain-making (by Kiana Labbaf Khaniki)

the names of 41 bald men all living in the village. They hang the paper from a tree branch and wait for the rain. In the village of Gir, Nishabur, this ceremony takes on a new dimension. In the time of drought, the villagers gather around someone who holds a long thread in his hand. The crowd names as many bald men as they remember, and the man ties a knot in the thread after calling each name until number 40. Afterward, they steal a pottery from a house and fill it up with water. They burn the knotted thread and put its ash in the pottery. Someone takes the pottery to a rooftop and pours the water into a gutter in the hope that it would rain (Bashguz 1979, p. 136). This ceremony turns more violent in the tribe of Bakhtiari. They perform a rite called "forty baldies", which has nothing to do with drought. In contrast, this rite is performed when rain does not abate and people fear floods. In this case, they gather 40 sticks, each of which is named after a bald man in the village. They look for those 40 bald men and catch them all, as if it is their fault that rain does not let up. They tie up all bald men and beat them hard, each with a stick specific to him. The villagers take the bald men hostage to get their families to pester God to make the rain stop. Hence, each bald man's senior son or daughter goes onto the rooftop and prays to God to stop the rain by singing the following song:

"Our forty bald men are tied up.

My God! please have a mercy on us and stop the rain.

His senior son looks pale.

The flood is rushing down" (Khazayi 2015, p. 323).

Sometimes, such ceremonies take on more violent dimension. In the villages of Kermanshah, western Iran, a rite called Gabruwa is performed, which is more of a horrific fighting. The villagers invade a neighboring village and plunder them to take whatever they find as booty, like their cows, sheep, chickens, and so on. The looted village struggles to get their livestock back, and as a result a real battle would be staged, in which both sides batter one another with heavy clubs. They believe that a fighting like that can shame God into sending rain down, since people fall out with one another owing to God's negligence toward them. If it rains, they would return their booty, otherwise they keep them hostage until it rains (Papoli Yazdi and Jalali 1999, p. 196).

There is a wide variety of such rites across Iran, whose ultimate aim is to disturb the social norms in order to spite the divinities who are believed to determine, rule and own the norms. For example, In Maragheh, Eastern Azerbaijan province, they climb up the rooftops and steal the gutters. In Minab, Hormozgan province, people go on a short strike and do not work on a Thursday. In Zanjan, Zanjan province, some people wash a donkey's head. Sometimes they gather human bones from a cemetery and burn them all, a sinful act that is subject to punishment in Islam. In Mazandaran and Gilan, they take a pulpit out of a mosque and put it in water until it rains. Sometimes they even take off a shrine's door and put it in water, as if they take it hostage (Zolghafari 2016, p. 86). They unconsciously believe that an act of defiance may make the divinities relent and loosen their grip on the rain, like modern people who riot against the government to force it to compromise (Fig. 6.6).

6.2.2 Supplication

In the village of Samia, Bushehr, raining in October and November is very crucial to the cultivation, and people would become apprehensive if it would not rain enough. Hence, people go to a place west of the village, named Qebleh Doa, where they can pray and arouse God's compassion. They go there in droves, taking with them a rooster, a hand-mill and a leather bottle inflated with their breath. A leather bottle containing air instead of water can better show their grievance. While walking toward the place, they sing together:

Fig. 6.6 Gabruwa ceremony in which a neighbor's cow is held hostage until it rains (by Kiana Labbaf Khaniki)

"Let's go to Qebleh Doa in the hope that God would have mercy on us".

On their way, they sing out the following song, while passing by a tree. They ask even the tree for help.

"You! meek green tree! please let the rain come down.

Our small watermill is in need of barley.

Our little leather bottle longs for water.

Our little chicken looks for birdseed".

But men took off their hats in frustration, and women start beating their own chests in supplication. Some of them go mad and hit the poor rooster and blow up the leather bottle. They then return to the village and parade along the alleys, while chanting:

"Lord! please look down!

Look at my hungry stomach.

Look at my thirsty heart.

Look at my bare feet.

Water, Water, Barley bread!

My little bottle longs for water.

My small hand-mill waits for barley.

My little chicken wants birdseed".

The crowd stops by each home on their way, asking for something like a handful of flour, peas, or even firewood. They repeat this ceremony several times until it rains. They put the foodstuff gathered during the ceremony in a big bowl to make a special soup, and also they drop a few coins in the soup. They give out the soup to the villagers, and those who happen to find a coin in their soup are believed to make a fortune in the future.

Sometimes local people look to a stone to bring them rain. For example, in the village of Khour in Birjand, there is big and heavy boulder in the village moat. They believe that sky would have mercy on them if they manage to roll the boulder over. All the villagers come together around the moat where the young men use some strong woods to lever the boulder, though it was very difficult and laborious to do so. If they fail to move the boulder, they come back next day and try their best until they succeed. When the boulder is rolled over, they wash it thoroughly and then place a smaller stone on its top. They believe that the big boulder has once been an old woman who has mysteriously petrified, and the small stone can be her child. When they finish with this job, they sit down and wait for the rain in the belief that it would rain in one week.

Such a belief in stones can be found in other parts of Iran but with different deviations. For example, in the village of Ayask, Ferdows, people resort to a big stone, when they were desperate for rain. A group of women bring home a relatively big stone, and they sit around it on the floor. The 36th chapter of the Quran—called Yāsin—should be read out 40 times, and each woman should recite it only once if they are 40 in number. After they finish reciting, they take the stone to a cistern named Howz-e Haji, and tie a rope around it and hang it from the middle of the cistern ceiling just above the water. They consider it a good omen if the stone slips out of the noose and falls into the water. In this case, they believe that their prayer has been accepted and it would rain soon. If the rain does not abate and a devastating flood is feared, the women ask someone to dive into the cistern to retrieve the stone. They believe that it continues raining as long as the stone is there in the water.

In the village of Giahdan on Qeshm Island, in the time of drought, people manage to collect 366 pebbles on which they write "May God Quench Our Thirst". Each of those stones would be given to a pious man so he would recite the verse of Hamd from Quran 366 times and then bless the stone by blowing on it. Afterwards, they get all the stones back and put them in a sack and drop them all into a water well. According to locals, this ritual is based on an old story when people and animals were starving to death in the wake of a persistent drought. No water remained in the ponds and nothing for people to eat. People sat together to find a solution to that disaster. One of the elderly came up with a great idea according to which 366 small stones were gathered and a sacred word was written on them. Each stone was blessed by a man of God and then all the stones were dropped into a well. After a short while, a gray cloud appeared in the sky, looming out of the sky in the direction of Mecca. It started raining and gradually it turned into a torrential rainfall. All the pools, wells and rivers became awash with water, but the rain did not let up. Hence, people feared that their properties would be flooded and washed away into the sea, if it continued raining. They asked a nimble man to dive into the well and bring the blessed stones back, which were deemed the culprit for that torrential rain. To their surprise, the man brought the stones back, but the rain did not abate. They realized that only one stone had been left behind, when they counted the stones that had been gathered by the diver. The diver again plunged into the water in search of the last stone. It stopped raining as soon as the diver showed up and put the last stone on the ground.

The relationship between stone and rain is traceable in Islamic mythology. It is said that Japheth, one of the three sons of Noah, decided to leave Suq al-thamānin[3] for eastern lands in Turkistan. In his farewell to Noah, Japheth begged his father for teaching him a prayer that he could utter when in need of rain. Noah blessed a stone and handed it to him. Japheth went to Turkistan and became the ancestor of all pastoral tribes there. He freely roamed all steppes of Turkistan with his flock, and invoked the same stone whenever he wanted the clouds to make rain. "Later, this stone was called Jedda Tāsh among Turks" (Abulfazl ibn Mubarak 2006, p. 99). That was a normal stone that had acquired its rain-making quality through Noah's blessing, but a piece of a meteorite has been believed to intrinsically have such a

[3]Suq al-thamānin is believed to be the first settlement that Noah established at the base of Mount Judi, the place where his Ark descended after the Great Flood (Haj Manoochehri 2010, p. 699).

magic ability. In many places, people make sacrifices for meteorites to solicit rain from it (Eliade 1993, p. 223).

In the village of Afin, Qayen, south Khurasan province, a group of kids make a wooden cross, put a hat on its top and put a ragged coat on it, in the shape of a scarecrow, which is called "Talu". They carry Talu around, singing a traditional song as follows:

"Our Talu is thirsty.
He wants holy water.
The shepherd longs for cheese.
The lamb craves for milk.
God! Please give us rain.
For the sake of saints
The lambs have eaten so much dirt that they are dying.
The calves have eaten so much dry straw that they are dying.
God! Please give us rain.
For the sake of saints
This lame dumb goat is moaning at the mouth of the dry well,
With sorrow for lack of rain and snow.
At the base of the mountain Laleh Zar, an army has camped.
Just in the hope of spring rain.
God! Please give us rain.
For the sake of saints
Molla Reza's cow has kneeled down before God
Molla Reza's cow has kneeled down in the middle of path.
The wheat is buried under soil.
It is suffering from thirst.
God! Please give us rain.
For the sake of saints"

The neighbors pour some water on Talu and then give foodstuff to the kids, in the hope that it rains. A very similar ceremony is held in many villages of south Khurasan province. They make a doll dressed in ragged clothes. The doll's skirt and veil resemble the local women's traditional clothes. The doll is called Atallu that is reminiscent of a local poor woman desperate for water. A group of youngsters carries the doll around the village and stops by every door to ask for permission to sing the song of Atallu. When the house-owner lets them sing, they chant together the following song:

"God! Please make it rain, as much as a waterfall!
Our poor Atallu is thirsty.
She wanders in the mountains in search of water.

She wants the water of Ghadir[4].

She wants a spoon of milk.

God! Please make it rain on the shepherds' sticks, on the farmers' shovels, on the herders' whips.

Shepherds want cheese and bread, their dogs want dough, their kids want milk.

Our yellow happy goats used to give birth to twins, but are dying of thirst this year.

Have a mercy on our sheep, on the ewes' manes, on the goats' horns.

Our wheat is under soil, desperate for water.

It would be great if it rains. The farmers would fall in love, if it rains" (Shateri et al. 2013, p. 99).

Every house gives them a handful of foodstuff with which they cook a special soup and give it out among the villagers (Fig. 6.7).

Some of the rain-claiming ceremonies have considerable artistic value. For example, in Torbat-Jam and Tayebad in Razavi Khurasan province, someone plays "Dutar", an indigenous long-necked two-stringed instrument, to attract the rainy could. They believe that singing the "song of rain" to a traditional music can gain the affection of sky and make it rain. There are some musicians called "Allah Baran" who specialize in playing this kind of music. At the time of drought, the villagers invite a musician to draw the cloud with the magic of his music. He sits on top of a hill, dressed in clean white clothes, and sings and plays as emotionally as possible for a couple of hours. The typical lyric sung on this occasion is as follows:

"God! Please give us rain and your blessing.

The red tulips are moaning from the pain of thirst.

They are looking forward to rain.

Please give us rain!

Those who are dry farmers have no water in their ditches.

They are looking to you.

Please give us rain!

Please send down rain, an endless rain.

The red tulips are moaning under soil.

Give us rain for the sake of shrines, for the sake of Koran verses.

[4]Ghadir was an area between Mecca and Medina in Saudi Arabia, where the caravans of Muslim pilgrims stopped by to drink from a pond named Ghadir. This pond was filled with rain water, which was a boon to the thirsty pilgrims crossing the desert. This pond is of great value to Shia Muslims, because they believe that at that place Prophet Mohammad proclaimed Ali his successor.

Fig. 6.7 Atallu—a doll dressed in ragged clothes—which is believed to be able to make rain (by Kiana Labbaf Khaniki)

Given us rain, all animals have fled to the mountains.

They are praying to God.

The white cloud has shied away.

But you! black cloud! Please shower down!

God! For the sake of Muhammad, please give us endless rain".

6.2.3 *Appeasement*

On Qeshm Island, locals believe that there are some sinister metaphysical creatures that stop it from raining. Hence, people try to appease them by offering delicious foods to them in the hope that they let go of the rain. A group of black men and women tie a piece of cloth around their waist and line up, while holding each other's waist with one hand and holding a basket with the other. They hang around this way and sing out a traditional song to invite the villagers to donate some foodstuff like wheat, barley, sugar, rice or even money. All the foodstuff are taken outside the village, where they cook a food with no salt. No one is allowed to taste the food that should be dumped on a clean ground for the sake of those creatures.

Appeasement toward metaphysical powers culminates in sacrificing animals. For example, in Andika, Masjed Soleiman, there are three special big stones called stones of rain, and someone named Seyyed Jalal is their custodian. The villagers wash the stones with pure water as ablution, and then move the stones in the direction of kiblah (direction that should be faced when a Muslim prays). Afterward they pray toward the stones, and then sacrifice a cow or a sheep whose blood should be rubbed on the stones. They take it for granted that it will rain after that ceremony, and they run to the stones and change their positions again as soon as it rains. Because they believe that the rain turns into a flood if the stones would be left in the direction of kiblah for a long time.

Sometimes they look to natural phenomena that are believed to have a mysterious connection with raining. For example, in Sabzevar, Razavi Khurasan province, the villagers make a special doll called "Chuli Chaqal", with a piece of rag and a stick. A large group of kids carries the doll down the street, singing the following song:

"Chuli Chaqal! Bring us rain.

Lower the price of wheat.

Wheat is hidden under soil.

It's dying of thirst."

The kids knock at every door, where they can get some wheat flour with which they make some small dough balls. They took the dough balls to the qanat exit, where they drop the balls in the water for the fish. This way they feed the qanat fish in the hope that fish's joy and satisfaction would solve the problem. The same ceremony is practiced in the village of Jilu, Nishapur, Razavi Khurasan province. In this village, an old woman is living, who is called Mother Fish. The woman takes care of the qanat fish, and protects them against those who try to catch or bother them.

6.2.4 Combination

In some regions, the villagers perform some ceremonies that represent appeasement, defiance and supplication at the same time. For example, in the village of Zohan, Southern Khurasan, people respond to a rainless season by going to the desert while wearing their clothes inside out. Any action against the social norms can be considered defiance against divine rules, because in the traditional communities all the social regulations are legitimized by God or metaphysical phenomena. They go to the desert in droves with their clothes inside out to pray to God for rain. In the same ceremony, they shift from defiance to supplication to ensure rain. If this ceremony does not work and rain does not show up, they stick a donkey's skull on the tip of a cane and put it on the peak of a hill. This act is akin to profaning sanctity of heaven, in the hope that God pays more attention to the people who are running out of mind. If this method does not work, they manage to destroy the crows' nests wherever they find them, in the belief that a crow might have stowed a human bone in its nest and the bone may belong to a heretic who has never prayed to God during his/her life. They pull down the nests one after another and search through them until they find a bone. They see the bone as the culprit that should be put on fire to break the spell, and then wait for the rain.

Another example is from the village of Feyz Abad in Surmaq, Fars province, where people pick a single young man, and dress him in ragged clothes. He drapes a shawl around his waist and hangs some small bells from his neck. Two of his friends walk along with him, while carrying two sacks for donations and gifts, and another man follows them while playing a frame drum. The young man waggles his body in an awkward way to sound the bells, and more and more people join and follow him from home to home. Each home gives him something like a handful of raisin, almond, walnut, sugar or a few coins, which are tucked in the sacks. Whoever donates pours water on the young man immediately after that. When the young man finish with this job, he heads to the qanat exit, accompanied by the crowd. The crowd goes for him and pushes him into the qanat water. They do not allow him to get out of water until someone intervenes and says: "please let go of him. On behalf of this young man, I promise that it would rain". The crowd asks: "when?" The surety replies: "just in ten days". Thus, the young man would be set free, but every day he goes to the village mosque and prays for rain. If he would be lucky enough, it would rain, and people no longer bother him, otherwise they catch him again and take him to the qanat water and plunge him into water until someone else stands surety for him. This ceremony may continue until it rains. The two elements of "tormenting an innocent young man" and "praying to God" are expressive of defiance and supplication as two cultural strategies to persuade the divine forces to send down the rain (Fig. 6.8).

In Azerbaijan, people start the rain-claiming ceremony with supplication but end with defiance. In spring around early May, they wrap a handkerchief around a wooden spoon in the shape of a doll. They call the doll "Chumchah Khatun" that literally means madam spoon. Someone takes the doll to the street and walks around along with a group of teens who sing the following song to a traditional drum:

> "What does Chumchah Khatun want?
>
> She wants rain from God.
>
> Her hand is stuck in dough.
>
> She wants a spoonful of water now.
>
> Clouds of mountain! Hope of orphans! Almighty God! Please send down the rain.
>
> Our wheat and barley are wilting in the thirsty farms.
>
> Girls! We are going to the garden to pick some flowers.
>
> We are going, while I am becoming a cloud and you are becoming rain.
>
> Let's go to the garden while raining.
>
> Please send us rain, so my tresses can grow longer.
>
> My mother has a shirt. Please send us rain and drench her shirt".

Every house gives some foodstuff to the group, and in the end they go up a hill and put a big pot on fire to cook a special soup with the foodstuff they have

Fig. 6.8 A young man dancing in the street to make rain (by Kiana Labbaf Khaniki)

collected during the ceremony. Everyone hopes that it would rain, while the pot is boiling. If it rains before the soup is ready, the group sit together and eat the soup with joy; otherwise they knock the pot over and pour the soup on the ground with anger (Khazayi 2015, p. 236).

6.3 Marriage of Qanat

One of the weirdest rites performed for qanats, is the marriage between an inanimate object and human. First of all, this custom is rooted in a dichotomy between male and female objects according to the villagers' mythology. Classifying inanimate objects is associated with "cognitive revolution" that occurred some 70,000 years ago (Harari 2015), when invention of grammatical languages enabled humans to personify their surrounding world. Gender was the most prominent and primitive distinction that our ancestors could make between the members of their human groups (Wheeler 1899, p. 529). Before they could differentiate between rich and poor, noble and commoner, believer and heretic, and so on, humans came to realize that all people fell into two groups; male and female, based on a difference that was related to one of their vital traits: sexuality. Therefore, our ancestors began personifying nonliving things by classifying them into male and female, the same as they did for their fellow humans. After tens of thousands of years, we are still using their funny word game in many languages around the world, in which even stones, trees, moon, sun and wind carry a gender, a peculiarity that is called grammatical gender by linguistics (Ibid.).

Although in modern Persian language there is no division between words according to their sex, ancient Persian was packed with a gender specific vocabulary (Hosseini and Ebnorasool 2009: 56). Classification of nonliving phenomena into male and female is reminiscent of their primeval approach that was once manifested in their language.

In many Iranian villages, people believe that some water sources are female whereas some others are male (Dadmohammadi 2015). For example, in Arak, Markazi province, the qanats in whose water the worker's skin becomes dry and develops xeroderma are regarded as male, otherwise they are called female. In Izadkhast, Fars province, the gender of water sources is recognizable by their sounds. If a qanat flows gently and quietly, it proves to be female. But if a qanat gurgles down, it is regarded as male. In the village of Garmab, Hamedan province, the qanat whose water fluctuates from year to year is regarded as male, whereas the qanat that enjoys a more stable water discharge is believed to be female. In Tuyserkan, Hamedan province, the qanat whose tunnel still enjoys groundwater seepage even downslope from its mother well, is called female (Farhadi 1985, p. 8). Sedaqat-Kish (2003, p. 35) ascribes this belief to an ancient Zoroastrian mythology according to which some waters are patronized by a male divinity named Tishtrya and some by a female one named Anahita. Attributing sexual identity to water bodies has been mentioned even in a well-known geography book written by al-Maqdisi (1982, p. 171) in the tenth century AD, where the author clearly writes: "the river Tigris is female,[5] since it is tender and loved by clergymen, whereas Euphrates is male[6]".

In the village of Hunjan, Isfahan province, there is a river named Zar-Cheshmeh that gurgles down a mountain with a loud noise, and scratches one's skin if they put their hands into this water. The villagers say that Zar-Cheshmeh in Hunjan is male. They believe that one cannot sleep close to a male water, but a gentle female water lulls you into a sweet sleep. According to them, the spring "Ab-e Kah" is male too, whereas the springs "Garro" and "Muk" in Izadkhast are female.

When Zar-Cheshmeh begins dwindling, people say that its water is looking for a female water in the neighboring village Izadkhast. They come together in front of the village chief's house to find a solution. They usually decide to single out an agile young man who can go to the village Izadkhast and steal a bucket of water from their female spring. This is a risky and adventurous mission, because the inhabitants of Izadkhast try their best to stop him from stealing their water. If he would be caught by them, his bucket would be confiscated and he would be beaten

[5]In Arabic: Ma' Onsā.

[6]In Arabic: Ma' Zakar.

up by an angry mob. People of Izadkhast say that a bucket of water taken from their female spring and added to a male spring can lead to a decline in the water of their spring, because their spring would become absent-minded and distracted as soon as it savors the experience of meeting a male water. That is why they bend over backward to prevent any contact between their spring and a male one. On the other hand, a man from Hunjan lurks near Izadkhast spring to fill up his bucket at the first opportunity.

The water thief stealthily goes up to the male spring of Izadkhast on a Thursday morning, while his fellow villagers follow him from afar. The water thief puts up a tent over the spring to cover it, so that no one even the sky can see him steal water. He plunges his canteen into the water to fill it up and flings it into his backpack. The water thief waits somewhere just outside the village of Izadkhast, until one of their people comes into sight. As soon as he sees a man from Izadkhast, he shouts out that "beware! I just stole your water", and then he runs away immediately. He should not put his canteen down on the ground or turns around to look back, while running. The men of Izadkhast chase him as fast as possible to catch him and get the water back, before he reaches his village Hunjan. If he manages to bring the stolen water to Hunjan, the village chief applauds him and sacrifices a sheep or even a cow in his honor. Afterward, the village chief along with other village men take the stolen water to their own spring and pour the water down into the flow in the hope that the spring would perk up and bring more water. The next day, they throw a party where lots of sweet cookies are handed out to the villagers to celebrate their successful water theft (Fadaee 2006, p. 176).

Such copulation does not take place only between male and female springs or qanats, but sometimes a human may become part of this story. A marriage between a woman and a qanat may seem akin to human sacrifice whose vestige has still remained in this way in the Iranian villages (Baghdasarian 2002, p. 31). Nonetheless, it does not seem very plausible to me that this kind of marriage is reminiscent of an era when humans were sacrificed for water bodies. No archeological evidence or historical record suggests that such a rite has been practiced in Iran at least since 3000 BC, and even animal sacrifice was denounced and banned by Zoroaster in the second millennium BC let alone human sacrifice (Aghakhani Bijani et al. 2018, p. 5). Apart from some sporadic conjectures, we do not know much about the cultural elements of the prehistoric communities that lived on the Iranian plateau tens of thousands of years ago. Even if they really practiced such a harrowing ceremony in which a woman was sacrificed for a water spring, the modern Iranians have as little in common with them as with Inuit people for example. It is very hard to believe that the modern Iranians have inherited the residues of a primeval culture and remained fixated on it even after tens of thousands of years. From anthropological point of view, Iran has always been a very dynamic land where different cultural systems have appeared on stage one after another over the past millennia. If we insist to trace back "qanat marriage" to those primitive groups, I prefer to attribute it to a primitive animism that is still living on even in our modern cultures. I will return to this notion after the following case study.

One of the best examples of such marriage has been reported from the village Alim-Abad in Arak, Markazi province. This village enjoys two qanats named Kahriz Bala and Kahriz Payin. According to the locals, the qanat of Kahriz Bala is a male qanat, since its water carries more minerals and it is hard for them to digest its water. The qanat of Kahriz Payin whose water is fresher and purer is regarded as female. Nevertheless, they never try to marry Kahriz Payin off to the male qanat of Kahriz Bala, but they believe that their male qanat gets more satisfaction from a marriage with humans. The last time the qanat of Kahriz Bala married a woman was 1971.

According to the villagers, a severe drought struck the village around 1963 and persisted until 1971, when many of their farms and orchards withered away and many of the inhabitants had to migrate to the cities. Their qanat was about to dry up, and its water dwindled to the extent that its water could not wet their thirsty farms. They did whatever they could to save the qanat. At the exit point of their qanat, they sacrificed a couple of sheep in the hope that more water would flow out in return. Several times, all the villagers went to the desert and held a mourning practice in order to beg Imam Hossein (one of the Shia saints) to intercede for them with God and stop that calamity. But none of those solutions worked, until they came to decide to marry one of the village women off to the qanat, as the last resort. They remembered that in the past the qanat enjoyed a happy marriage with a woman named Mah-Taban who eventually died around 1958 and left the qanat a lonely widower. They realized that the only thing amiss was a wife that they should have found for the qanat.

The villagers nominated a woman called "Aunt Zobeideh" whose husband had recently passed away. The qanat's future wife should have lived up to a few qualifications of which being a needy widow was very important. Eventually, a group of villagers along with the village Mullah (Muslim priest) came up to Aunt Zobeideh to propose to her on the qanat's behalf. Aunt Zobeideh accepted the qanat's proposal, so everyone sat together to negotiate about other details like the amount of her dowry and allowance. Her allowance was agreed to be shared among the villagers on the qanat's behalf. In return, Aunt Zobeideh accepted to bath in the qanat water at least once a week to do the qanat's heart good (Safinejad 1980).

The villagers celebrated this wedding, and the next day Aunt Zobeideh privately went to the exit of qanat to consummate her marriage by getting naked and sitting in the water flow. On the same day when she slept with the qanat that way for the first time, she was given 300 kg of wheat. Afterward everyone in the village was obliged to give her between 15 and 30 kg of wheat twice a year, one time in the month of Ramadan and again in the month of Muharram, which are both revered by Shia Muslims. The qanat water started increasing shortly thereafter, according to the locals.

The villagers felt indebted to Aunt Zobeideh for saving their qanat from desiccation and for such an upturn in their cultivation. Everyone treated her well, and never withheld their favors from her. Aunt Zobeideh lived an easy life until 1981 when she lost her sight and became totally blind. Nonetheless, she still groped her way to the qanat water to fulfill her duty that was believed to keep the water flowing. In 1986, her marriage ran out of steam and Aunt Zobeideh wanted a divorce, since the allowance could not keep up with the rising inflation. She had to pay in cash for the goods she needed, whereas her allowance was paid in kind as sacks of wheat. Her husband had not updated his financial affairs that were foreign to the modern economic relations of the village, in which the Ministry of Agriculture monopolized the purchase of their wheat harvest, and the farmers could spend this money on whatever they needed including bread in the village market. The village traditional water mills and the home bakeries fell out of favor, and there was no longer point in storing wheat at home, asking the miller to grind a sack of wheat for her, stoking the oven and managing to bake breads. There was no watermill left to grind Aunt Zobeideh's wheat, and besides she wanted to enjoy a better life in which more various products could be consumed through the new monetary system.

When the villagers faced Aunt Zobeideh's insistence on divorce, they tried to change her mind by offering some money instead of wheat. Aunt Zobeideh was reconciled with the qanat over her last years, when she received a diversified stipend including money as well as tea, sugar, clothes, and so on. Although the qanat of Kahriz Bala eventually gave into the socio-economic transition in the village by upgrading its financial system, his wife did not last any longer and eventually died in 1992 (Fadaee 2006, p. 174). She was the qanat's last wife, and after Aunt Zobeideh her fellow villagers lost their belief in qanat marriage at all (Fig. 6.9).

Marriage of qanat was practiced in many places like the village of Bakhsh-Abad in Damqan, Semnan province (Farhadi 1997, p. 177; Ashtiani 2004, p. 11), Tafresh in Markazi province (Tafreshi 1982: 58), the village of Hosseincheh in Isfahan province (Safinejad 1980, p. 243), the village of Ben in Chaharmahal and Bakhtiari province, and Delijan in Isfahan province (Fadaee 2006, p. 175).

In many places, this cultural rite had an economic function as well, which benefited the qanat in a more realistic way. Qanat cost is not limited to the money spent on its construction, but it continues into its operation, because a typical qanat needs repair and cleaning once in a while in order to keep running. Sometimes, the qanat marriage was subtly associated with financial affairs of qanat. In some regions, the locals used to hold a wedding ceremony for their qanat only when it came to dry up. The villagers treated this wedding and the qanat's wife with the same attitude they had toward a real bride, and all the wedding customs were observed exactly like a real one. One of those customs was the wife's allowance, and the onus was on the husband to pay it according to their culture. Given that the qanat could not fulfill its commitment, the village people raised an amount of money once in a while to pay the bride on her husband's behalf. Of course, the money collected this way was more than what the qanat's wife really needed, so the rest went to the qanat maintenance and cleaning. Thus, those people were right in their belief that marrying off their qanat can increase its water. Although this custom has long been abolished, there is still its vestige in many Iranian villages. The allowance paid to a wife is called "Nafaghe" in Persian language but this term also now refers to the money that villagers collect to clean and repair their qanat. The name of that ancient custom is living on, though there is no longer a trace of it in reality (Figs. 6.10 and 6.11).

Fig. 6.9 Aunt Zobeideh the last qanat's wife in Iran (Safinejad 1980)

قبض رسید نفقه حسین آباد رستاق یزد

تاریخ ماه ۱۳ نمره

مبلغ ریال توسط بابت جره ملکی

جهت حفر کار جدید در سال ۱۳۱۵ واصل و عاید گردید

محل امضاء مباشر نفقه محل امضاء ناظر

Fig. 6.10 A qanat bill called Nafaghe issued to collect the shareholders' contributions, in the qanat of Hossein-Abad Rostaq, Yazd province

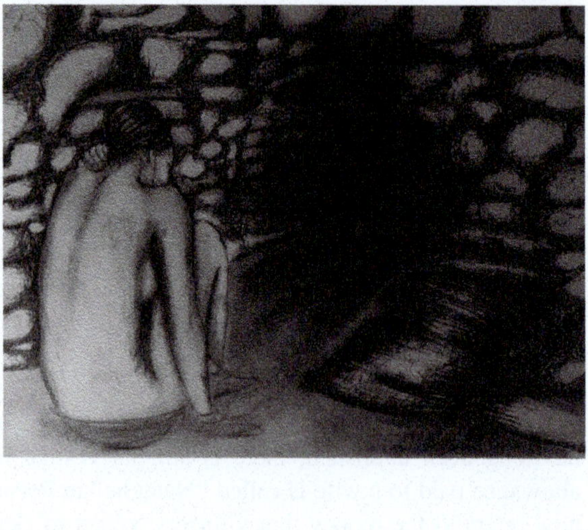

Fig. 6.11 Qanat's wife bathing in her husband's current (by Kiana Labbaf Khaniki)

6.4 Conclusion

In Iran, "water" has always had the power of juxtaposing different spaces and locations that are incompatible with each other, a power that can create different types of heterotopias (Foucault 1997). Water is the main element of an Iranian religious place as a heterotopia, where different spaces overlap to ensure purification, sacredness, satisfaction and immortality. Another example is a Persian garden with its irrigation system, which brings together disparate scenes from the sacred cosmology, as heterotopia of illusion according to Foucault (Sudradjat 2012, p. 31). Heterotopia of water is dynamic by its nature, absorbing more cultural elements in the course of time. Any cultural element can be potentially deployed in creating this heterotopia swallowing whatever trapped in its strong gravity like a cultural black hole. Āb-Anbār well exemplifies the dynamic nature of "hydro-heterotopia". Āb-Anbār is a traditional reservoir that stores drinking water for a particular residential area, being filled by a qanat during the winter. Āb-Anbār is simply a water tank from which thirsty people can drink, but it has taken on different incompatible dimensions most of which seem irrelevant to its initial biological purpose, to the extent that we can say that Āb-Anbār has been turned into a heterotopia. A high level of architectural skill is deployed in building the Āb-Anbār whose decorated entrance, beautiful huge dome, and minaret-like wind catchers are all reminiscent of a mysterious sacredness typical of a mosque. Āb-Anbār dictates the urban layout, where almost all main alleys gravitate toward it at the center of a district in order to facilitate the residents' access to drinking water. The upper class tend to use their wealth and authority to settle around Āb-Anbār, which ensures their easy access to clean water, whereas the lower class occupy the spots further away from the same water source. Hence, social stratification is manifested in Āb-Anbār and its urban hinterland. Āb-Anbār is built and maintained by affluent people as acts of charity, but in fact they do penance by giving part of their surplus wealth to the public in order to dampen social tension. Āb-Anbār may have two separate stairways of which one belongs to Muslim and the other is specific to Zoroastrians in order to observe religious segregation even at the time of drinking water. Āb-Anbār serves as a congregation place where people exchange their ideas, dreams and myths, some of which constantly add to this growing hydro-heterotopia in the course of time. At a glance, Āb-Anbār is only a water tank, but it actually embodies the sanctity of water, social conflicts and bonds, and religious ideologies, in the form of a dynamic hydro-heterotopia that encompasses all aspects of the inhabitants' social and cultural life.

Thus, all rain-making ceremonies invoke different incompatible spaces through a complicated system of symbolism. Sometimes, a heterotopia is devised to indicate that there is something amiss in their social norms, where social norms are mostly anchored in the religious moral codes. A heterotopia of this kind is constituted by a set of incompatible social behaviors to shame God or any omnipotent divinity into fixing the chaos by putting the only missing piece back in the jigsaw puzzle of the normal social life, which is "rain". A rain-making ceremony may display a surreal scene in which some cultural phenomena are at play in an awkward way. A male bride dressed in outlandish clothes, singing strange nonsense songs, taboo behaviors like teasing holy men or disrespecting holy places, denying traditional dress codes, and so on are more of an anarchic dream that forms and exists in parallel with reality. This heterotopia is intended to trespass on the domain of "reality" which is believed to be created, handled and monopolized by the deities, in order to move them to fix all problems that have befallen "reality" including the lack of rain (Fig. 6.12).

As for the qanat marriage, its economic function may explain its perpetuation rather than its origin. Qanat marriage seems to be rooted in animism or anthropomorphism that is one of our most primitive cognitions but still present and widespread in our cultures. According to Guthrie (1993, p. 64), animism or anthropomorphism is normal, because it results from a strategy universal in human perception, and it helps us understand perceptual and cognitive phenomena.

Animism set the stage for the earliest forms of religion (Praet 2014, p. 2) as a product of high-level abstract thinking (Willerslev 2018, p. ix). Animism is defined as an idea of pervading life and will in nature, attributing the human properties to the other elements of nature. Animists have in common that they impute life and/or humanity to this, that, and the other in a remarkably generous manner (Praet 2014, p. 3). Animism started as a philosophy of human life, and then expanded itself until it became a philosophy of nature at large (Tylor 1871, p. 169). With that early philosophy, human life seems the direct key to the understanding of nature in general (Ibid., p. 99).

Animism has never died out, but it has underlain many of our philosophical attempts to perceive and explain ourselves and the world we are living in. Some vestiges of animism have survived even in our today religions, like the genes of our Australopithecus ancestors of which we cannot get rid. According to Nalder (1926, p. 77), Islamic ablution is to free the worshiper not from bodily but from demoniac and supernatural pollution. In Islam, gargling with the same water they use for ablution can bring them God's blessing and keep Satan away from them (Zahedi

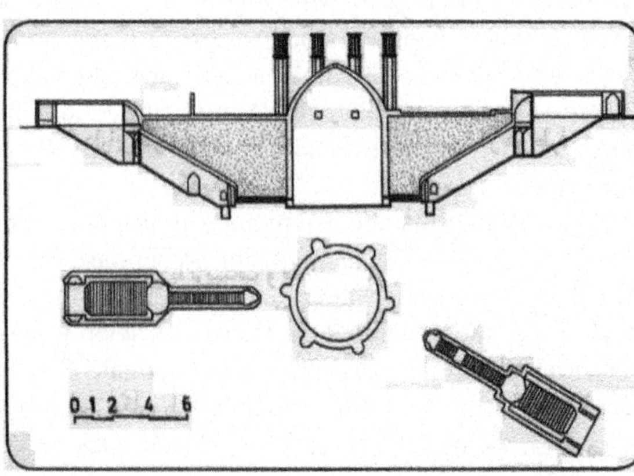

Fig. 6.12 Āb-Anbār of Shesh-Badgiri in Yazd, with two separate stairways one for Muslims and the other for Zoroastrians, both leading to the same water tank (Maserrat 2010, p. 250)

1998, p. 56). According to some Islamic scriptures, natural phenomena are not indifferent to people's deeds, but human sins can provoke different reactions from them. For example, adultery can cause earthquake, and oppression of the weak and defenseless people shoos the clouds away and brings about drought (Amoli 2012, p. 720).

Animism has made it even to our modern minds, despite the strong barriers that our scientific education has ever built against such superstitions. We are amazed that indigenous illiterate people treat a water stream as a human who can even marry a village woman. But even nowadays among the well-educated people, you may meet some who believe that water is as intelligent as humans. They believe that water can listen to their prayers, and prayer can affect the water molecules by beautifying their geometrical figures. They even take a further step by saying that water can listen and react to the bad and good words in every language. "Water understands good and bad words. Water makes distinction between beautiful and ugly. Water listens to our prayers. Water reacts to insults and name-calling. Water is able to distinguish between positive and negative minds, between positive and negative writings, between good and bad behaviors. Water consciously reacts to these all" (Najjar and Vaziri 2015).

Animism is not limited only to the primitive cultures, but it has made its way even to our modern legislations in the developed countries. The Whanganui River is a good example, which originates from the center of New Zealand's North Island, running to the Tasman Sea. Since 1873, local people have struggled to obtain legal rights for this river that has been regarded as a real person according to their mythology. Eventually, as part of settling those ongoing disputes, a new act was passed in March 2017, which grants legal personhood status to the Whanganui River, acknowledging the river as a living whole that enjoys both physical and metaphysical elements. The act recognizes the river as a legal entity with all the rights, powers, duties and liabilities of a legal person. The river is represented by a guardian who is required to act and speak to the benefit of the river's health and well-being (O'Donnell and Talbot-Jones 2018, p. 4).

Developmental psychologists tend to link animistic behaviors to the concept of life that evolves ontogenetically through a series of stages. According to them, in these stages, life is initially attributed to anything that moves or resembles moving objects (Berzonsky 1974, p. 786). Hence, at early ages we are all subject to animistic behaviors that manifest themselves in speaking to the plastic dolls or believing in such fictional characters as Pinocchio the wooden animate puppet. This notion characterizes animism as a disorder, if a person would not move on, and remain fixated on this particular stage of their mental growth.

However, evolutionary psychology places a functional value on animistic ideas and behaviors, rather than sees them simply as mental disorder. Humans set foot in an ambiguous world that is awash with uncertainties. As human—especially at the early stages of our evolution—we more interpreted the phenomena in our environment than saw them (Guthrie 1993). Interpretation was the most important means for our ancient ancestors to perceive their world. Ambiguity that made humans resort to interpretation was mostly rooted in two factors: limitation of human senses that failed to detect everything in nature and deception that provided human with misinformation.

The primitive humans in their natural environment did not possess infrared vision to spot a leopard lurking near their cave at night. But they could interpret the rustle of branches and the sound of crunching leaves at a breezeless night. They barely noticed an eagle that swooped down from the sky out of the blue at such a high speed. But they could interpret the hectic forest and shrieking birds as indications that an eagle was around the corner. They could not smell a dreadful bear lolling in the dark corner of their cave, before entering unknowingly. But they could interpret the marks on the muddy ground at the cave entrance, which resembled a bear's footprints. Interpretation helped humans to survive those predators, and track down and catch the games. However, interpretation did not always turn out to be correct. Sometimes the rustle of leaves had nothing to do with a hungry leopard, but it was a ripe fruit falling down through the branches. Nonetheless, the advantages of their interpretation outweighed its disadvantages, and their ability to interpret was selected and accumulated in the course of their evolution over the past six million years. Their ability to interpret was sometimes misfired, when they encountered something like a microbe that was quite out of the reach of human senses but could inflict grievous impact on them.

In 1857, cholera became endemic in many Iranian towns and villages, and decimated a large portion of the Iranian population. At the time, Iranians misinterpreted the actual cause of this deadly disease, because this calamity on such a large scale could be afforded only by a very strong gigantic creature that was invisible like an evil ghost. They never suspected that all those destructions were done by a very small bacterium that was only 0.3 micron in diameter, based on their earlier experiences about their environment. Hence, they were sure that a crazy demon was rampaging through the streets to massacre them. Sometimes, angry men took to the streets, wielding sharp swords and shouting that "you damn disease, come to light if you dare! you can't stand my sword. I'm going to kill you". They even went to the desert in droves just outside the city, with their artillery. They fired many cannons at the same time toward the mountains while they blew many big horns and shouted as loudly as possible at the same time in order to scare the disease away (Nategh 1977, p. 55).

Besides, misinformation was another factor that made the world more ambiguous for the primitive humans. For example, the animals capable of camouflaging and even disguising as other animals or objects could deceive our ancestors. At sunset, it looked like sun was sinking into a swamp of red mud. When they looked up at the sky, they did not know that stars twinkled because of turbulence in the atmosphere of the earth, but they thought that twinkling was the way stars laughed. In Australia, Gamilaraay people believe that twinkling Venus is an old man who has told a rude joke and has been laughing ever since that way (Parker 1905). When our ancestors came across such phenomena, they spontaneously modeled them on things more familiar to them, like a laughing old man.

Therefore, limitation of human senses and misinformation through deception pushed our ancestors to use interpretation as their main tool to perceive their surrounding world. When it came to interpretation, they had a strong tendency to liken inanimate objects to something more familiar, whose behavior is more understandable and predictable. In fact, the only thing that human could really see was himself, and he had to interpret other things through his self-cognition which was a benchmark for perceiving the whole world. Thus, the ancient Persians interpreted sun as a man-like divinity who journeyed across the sky every day from east to west in a beautiful chariot (Bakhtourtash 1972, p. 32). In the same way, the Iranian farmers viewed a drying qanat as a single man who was pining away out of his passion for a woman. They prescribed the same remedy for the qanat as they did for a man who craved for marriage.

References

Abbasi, S., Safari, M., & Asgari, L. (2017). Kārkard-hāye Ostoore-shenākhti-ye Āb Dar Irān-e Bāstān [Mythological functions of water in ancient Iran]. *Āmoozesh-e Mohit-e Zist Va Towse'e-ye Pāydār, 6*(2), 117–126.

Abulfazl ibn Mubarak. (2006). Akbarnāme; Tārikh-e Gurkāniyān-e Hend [Book of Akbar; history of mughal empire in India]. In G. R. Tabātabāyi Majd (Ed.). Tehran: Anjoman-e Āsār Va Mafākher-e Farhangi.

Aghakhani Bijani, M., Toghyani, E., & Mohammadi, M. (2018). Āyin-e Ghorbāni Dar Shāhnāmeh-ye Ferdowsi [Ceremonial Sacrifice in Ferdowsi's Poetry]. *Farhang Va Adabiyāt-e Āmmeh, 6*(20), 1–21.

Aliabadi, M. (2012). Barresi-ye Anāsor-e Segāne-ye Nemādin Dar Jām-e Shoosh [A Study on the Triple Symbolic Elements on Susa Potteries]. *Honar-hāye Zibā, 17*(3), 5–12.

al-Jāhiz, A. B. (1920). Ketāb al-Hayawān [Book of the animals]. (Vol. 4). A. M. Hāroon (Ed.). Beirut.

al-Maqdisi, A. M. (1982). Ahsan al-Taqāsim Fi Maʿrefat al-Aqālim [The best divisions in the knowledge of the regions] (Vol. 1, A. N. Monzavi, Trans.). Tehran: Sherkat-e Moalefān va Motarjemān-e Irān.

Amoli, J. A. (2012). *Mafātih al-Hayāt [Doors to the life]*. Qom: Esra.

Ashtiani, M. A. (2004). Be Aroosi-ye Qanāt Davat Shodid [You are invited to the qanat wedding]. *Biyaban, 11*(530).

Baghdasarian, E. (2002). *Tārikh Va Farhang-e Kamarah [History and culture of kamarah]*. Tehran: Ehsān.

Bahar, M. (1998). *Az Ostooreh Tā Tārikh [From mythology to history]*. Tehran: Cheshmeh Publication.

Bakhtourtash, N. (1972). Gardoone-ye Khorshid Yā Gardoone-ye Mehr [Chariot of sun or chariot of mithra]. *Barresi-hāye Tārikhi, 7*(3), 1–33.

Bashguz, I. (1979). Rain Rituals and Rain-making in Iran (B. Farrokhi, Trans.). Ketab-e Jomeh, 18 (1979), 122–123.

Berzonsky, M. D. (1974). Reflectivity, internality, and animistic thinking. *Child Development, 45* (3), 785–789.

Boyce, M. (1985). ĀB: The concept of water in ancient iranian culture. *Encyclopedia Iranica* (Vol. 1). Routledge.

Carnoy, A. J. (2004). Iranian mythology (A. Tabatabayi, Trans.). Tehran: Elmi & Farhangi Publication.

Dadmohammadi, K. (2015). Tasir-e Qanat Bar Sākhtār-e Ejtemāyi Va Farhangi-ye Jamea'e Iran [Impact of qanat on the social and cultural structures of the iranian society]. In *Proceedings of*

the International Conference on Architecture, Urbanism, Development, Art and Environment; Future Horizon and a Glance at History. Iran.
Eliade, M. (1993). Resāleh Dar Tārikh-e Adyān [A history of religious ideas] (Vol. 1, J. Sattāri, Trans.). Tehran: Soroosh.
Eyvazi, R. (1966). Mazāher-e Shar Dar Āyin-e Zartoshti [Manifestations of evil in zoroastrian faith]. *Zabān Va Adab-e Fārsi, 18*(78), 152–171.
Fadaee, A. (2006). *Tamannā-ye Bārān [Desire for rain]*. Mashhad: Āstān Qods-e Razavi.
Farhadi, M. (1985). Aroosi-ye Āb [Wedding of water]. *Abziān, 4*(1985), 7–13.
Farhadi, M. (1997). *Farhang-e Yārigari Dar Irān [Culture of cooperation in Iran]* (Vol. 1). Tehran: Markaz-e Nashr-e Dāneshgāhi.
Foucault, M. (1997). Of other spaces: Utopias and heterotopias. In Neil Leach (Ed.), *Rethinking architecture: A reader in cultural theory* (pp. 330–336). NYC: Routledge.
Ghaemi, F., Yahaghi, M. J., & Poor-Khaleghi, M. (2009). Tahlil-e Nemādinegi-ye Anāsor-e Khāk Va Bād Dar Asātir Va Shāhnāme-ye Ferdowsi Bar Asās-e Naghd-e Ostooreh [An analysis on the symbolization of soil and air in Ferdowsi's poetry and mythology based on mythological critique]. *Adab Pajoohi Gilān, 3*(10), 58–82.
Ghazanfari, P., & Karimi, M. (2016). Cheshme Ali-ye Dāmghān Va Elal-e Taghaddos-e Ān [Ali spring in Damghan and the reasons behind its sacredness]. *Bāgh-e Nazar, 13*(42), 57–66.
Guthrie, S. E. (1993). *Faces in the clouds; a new theory of religion*. England: Oxford University Press.
Haj Manoochehri, F. (2010). Judi Kooh [Mount Judi]. *Dāerat al-Ma'āref-e Bozorg-e Eslāmi* (Vol. 18, pp. 699–701). Tehran: Markaz-e Dāerat al-Ma'āref-e Bozorg-e Eslāmi.
Harari, Y. N. (2015). *Sapiens: A brief history of humankind*. New York: Harper.
Homayooni, S. (1992). *Farhang-e Mardom-e Sarvestān [People's culture in Sarvestan]*. Mashhad: Beh Nashr Publication.
Hosseini, S. M. H., & Ebnorasool, S. M. R. (2009). Derangi Dar Pasvand-e "he" Va Neshāne-ye Taa'nis [A glance at the suffix "eh" as female indication in Persian]. *University of Semnan, Majale-ye Dāneshkade-ye Oloom-e Ensāni Vije-ye Adabiyāt-e Fārsi, 8*(28), 55–65.
Javadi, S. (2013). Bāz-khāni-ye Revāyat-e Bāstāni-ye Āb Va Derakht Dar Dowrān-e Eslāmi-ye Irān [A review on the ancient story of water and tree in the islamic period in Iran]. *Honar Va Tamaddon-e Shargh, 1*(1), 43–50.
Khazayi, H. R. (2015). *Afsāne-ye Bārān [The legend of rain]*. Mashhad: Māh Jān.
Maserrat, H. (2010). *Āb-Anbār-hāye Shahr-e Yazd [Water reservoirs of the city of Yazd]*. Tehran: Yazdā.
Masoomi, G. (1970). Naghsh-e Boz-e Koohi Bar Roo-ye Sofāl-hāye Pish Az Tārikh-e Irān [The figure of goat on the pre-historic potteries of Iran]. *Barresi-hāye Tārikhi, 5*(4), 33–68.
Mirnia, S. A. (1989). *Farhang-e Mardom [People's culture]*. Tehran: Pārsā Publication.
Moshiri, J. (2014). Marāsem-e Talab-e Bārān Dar Howze-ye Kaviri-ye Irān [Rain-making ceremonies in the desert areas of Iran]. *Ensān-Shenāsi, 11*(19), 118–138.
Najjar, V., & Vaziri, Z. (2015). Hooshmandi-ye Molkool-hāye Āb [Intelligence of water molecules]. Iran: Sama Organization.
Nalder, L. F. (1926). The influence of animism in Islam. *Sudan Notes and Records, University of Khartoum, 9*(1), 75–87.
Nategh, H. (1977). Tasir-e Ejtemāyi Va Eghtesādi-ye Bimāri-ye Vabā Dar Dore-ye Ghājār [Socio-economic impact of cholera in the period of Qajar]. *Majale-ye Tārikh, University of Tehran, 2*(1977), 30–62.
Nusservanji Dhalla, M. (1938). *History of Zoroastrianism*. London: Oxford University Press.
O'Donnell, E. L., & Talbot-Jones, J. (2018). Creating legal rights for rivers: Lessons from Australia, New Zealand, and India. *Ecology and Society, 23*(1), 7. https://doi.org/10.5751/ES-09854-230107.
Papoli Yazdi, M. H., & Jalali, A. (1999). Āyin-hāye Bārān-khāhi Dar Zamān-e Khoshksāli-hā [Ceremonies of rain-making at the time of drought]. *Faslnāme-ye Tahghighāt-e Joghrāfiyāyi, 54&55*(1999), 186–211.

Parker, K. L. (1905). *The Euahlayi tribe a study of Aboriginal life in Australia*. London: Archibald Constable.

Praet, I. (2014). *Animism and the question of life*. New York: Routledge.

Rasooli, J. (2015). Jān-e Zendegi: Negāhi Be Bāvar-hā Va Nemād-shenāsi-ye Āb [The essence of life: A glance at the beliefs and symbolism of water]. In *Proceedings of Iran Irrigation and Drainage National Conference*. Tehran: Irrigation and Drainage Association.

Safinejad, J. (1980). *Nezām-hāye Ābyāri-ye Sonnati Dar Irān [Traditional irrigation systems in Iran]* (Vol. 1). Tehran: Moasese-ye Motālea'at Va tahghighāt-e Ejtemāyi.

Sedaqat-Kish, J. (2003). Qanat-hāye Moghadas-e Irān [The sacred qanats of Iran]. *Ketāb-e Māh-e Honar, 57&58*(2003), 34–42.

Seddiq, M. (1975). Ādāb Va Rosoom-e Talab-e Bārān Dar Roostā-ye Samiyeh [Customs and rituals of rain claiming in the village of Samiyeh]. *Majale-ye Honar Va Mardom, 155*(1975), 55–57.

Shateri, M., Rajabi, M., & Rajabi, N. (2013). *Farhang-e Āb Dar Jonoob-e Khorāsān [Water culture in South Khurasan]*. Tehran: Fekr-e Bekr.

Stausberg, M., & Vevaina, Y. S. (2015). *The Wiley Blackwell Companion to Zoroastrianism*. John Wiley & Sons, Ltd.

Sudradjat, I. (2012). Foucault, the other spaces, and human behaviour. *Procedia - Social and Behavioral Sciences, 36*(2012), 28–34.

Tafreshi, M. S. Q. (1982). *Seyr-e Kootāhi Dar Joghrāfiyā-ye Tārikhi-ye Tafresh Va Āshtiyān [A short review on the geography of Tafresh and Ashtian]*. Tehran: Amir-Kabir Publication.

Tavakoli, M. R. (1975). Joghrāfiyā Va Tārikh-e Bāneh-ye Kordestān [Geography and history of Baneh in Kurdistan] (Vol. 1). Tehran.

Tylor, E. B. (1871). *Primitive culture; researches into the development of mythology, philosophy, religion, art, and custom* (Vol. 2). London: J. Murray.

Wheeler, B. I. (1899). The origin of grammatical gender. *The Journal of Germanic Philology, 2* (4), 528–545.

Willerslev, R. (2018). The anthropology of ontology meets the writing culture debate—is reconciliation possible? In K. Swancutt & M. Mazard (Eds.), *Animism beyond the soul* (pp. vi–x). New York: Berghahn Books.

Zahedi, Q. A. (1998). *Dastoorāt-e Teb Va Behdāsht Az Nazar-e Eslām [Hygiene and medical instructions from islam's standpoint]*. Qom: Hāzeq.

Zolghafari, H. (2016). Barresri Va Tahlil-e Namāyesh-hāye Bārān-khāhi Va Bārān-khāni Dar Adabiyāt-e Āmme-ye Irān [An study and analysis on the rain-making and rain-singing ceremonies in folkloric literature in Iran]. *Kohan-nāme-ye Adab-e Pārsi, 7*(4), 81–109.

Index

M. Labbaf Khaneiki, *Cultural Dynamics of Water in Iranian Civilization*,
https://doi.org/10.1007/978-3-030-58900-4

The manufacturer's authorised representative in the EU is Springer Nature Customer Service Centre GmbH, Europaplatz 3, 69115 Heidelberg, Germany. If you have any concerns regarding our products, please contact ProductSafety@springernature.com

Printed and bound by CPI Group (UK) Ltd, Croydon, CR0 4YY
22/09/2026
02221065-0008